CW00967653

ROCKET AND JET AIRCRAFT OF THE THIRD REICH

ROCKET AND JET AIRCRAFT OF THE THIRD REICH

TERRY C TREADWELL

SPELLMOUNT

For Wendy, Toby, Samantha and Rex.

Front of jacket: The rocket-powered Bachen Ba 349A, first flown under tow as an unpowered glider from a Heinkel He 111 in November 1944.

Back of jacket: Three-view of a Junkers Ju 287 and the remains of Johannes Steinhoff's Me 262 engine after crashing on takeoff. Steinhoff survived. See page 173.

First published 2011 by Spellmount, an imprint of
The History Press
The Mill, Brimscombe Port
Stroud, Gloucestershire, GL5 2QG
www.thehistorypress.co.uk

All images from the author's collection unless otherwise specified.

British Library Cataloguing in Publication Data.
A catalogue record for this book is available from the British Library.

ISBN 978 0 7524 6109 0

Typesetting and origination by The History Press
Printed in Great Britain
Manufacturing managed by Jellyfish Print Solutions Ltd

CONTENTS

Introduction 7

1 Messerschmitt Me 163B 'Komet' 19
2 Bachem BA 394 71
3 Fieseler Fi 103R Reichenberg 87
4 Messerschmitt Me 328 95
5 Heinkel He 178 and 280 99
6 Heinkel He 162 109
7 Henschel Hs 132A 123
8 Arado Ar 234 127
9 Junkers Ju 287 141
10 Messerschmitt Me 262 147

Specifications 179
Translations/Glossary 185
Index 187

INTRODUCTION

In A.D.1232, during the Mongol siege of the city Kai-fung-fu, the first recorded use of a rocket was identified in the Chinese chronicle *T-hung-lian-kang-mu*. The 'Arrow of Flying Fire' as it was called, was most likely to have been a hollow arrow that was stuffed full of some kind of incendiary material, or an arrow that had been attached to a similar substance.

In Europe around 1249, the English monk Roger Bacon (1214–1294) is credited with developing the first rockets in the Western world. Bacon was a philosopher and scientist and foresaw the use of gunpowder, together with mechanical cars, boats and aircraft. It is also said that he invented the magnifying lens but because of his scientific beliefs, the church branded him a heretic and in 1253 imprisoned him. He wrote a number of books during his time in prison and died in 1294, two years after his release.

In Russia virtually no records exist of the use of rockets until the 1600s, when documents accumulated by the Russian gunsmith Onisin Mikhailov were complied into a document entitled 'Code of Military, Artillery and Other Matters Pertaining to the Science of Warfare'. In this document detailed descriptions of rockets referred to as 'cannon balls which run and burn' are contained. Mikhailov's authorship of this early document has been disputed, mainly because the documents which made up the bulk of the manuscript were not entirely his, but a collection of some 663 snatches of information and articles from a variety of foreign military books and sources. But that doesn't really matter, the fact that Mikhailov collected these articles is of more importance than the debate as to whether or not he was justified in calling them his own.

Peter the Great of Russia devoted a lifetime to building Russia's military might. In 1680 he founded the first Rocket Works in Moscow. There they made illuminating and signal rockets for the army, under the guidance of English, Scottish, Dutch, German and French officers, who instructed them in their use. In the early 1700s Peter the Great made St. Petersburg the new capital of Russia and moved the entire Rocket Works to this new location, expanding it at the same time.

Germany's interest in rockets started seriously in 1923, when a German-speaking Transylvanian schoolteacher Hermann Oberth, published *Die Rakete zu den Planetenräumen* ('The Rocket into Interplanetary Space'). The book attracted a number of similar minded enthusiasts and in the backroom of a restaurant in Breslau called the Golden Sceptre, they met. At this meeting the society Verein für Raumschiffahrt (VfR) (Society for Space Travel)

Above: The Espenlaub Rak 3 showing the burnt section of the vertical stabiliser after its first flight. A contemporary of Alexander Lippisch, Gottlob Espenlaub is a now largely forgotten pioneer. His first aircraft was built in 1927 with the intention of working up to rocket-powered flight.

Right: Storch I.

Below: Lippisch Storch glider making its first flight.

Messerschmitt Me 163B Komet about to take off.

Campini Caproni in flight. This was the second jet aircraft ever to fly, at Milan on 27 August 1940.

was formed, the aims of which were to promote the theory of space travel and carry out experiments in rocket propulsion. Within two years the society's membership had grown from a mere handful of enthusiasts to 870 dedicated members. Among these was a young nineteen-year old by the name of Werner von Braun.

In 1929, an opportunity to publicise the society came when the famous German film director Fritz Lang wanted to make a film called *Frau in Mond* ('Girl on the Moon'). To launch the film Lang wanted Oberth to build a rocket and launch it to a height of 70 kilometres for the premiere. The problem was that Oberth was a theorist and not an engineer. So Oberth

The Horten flying wing that was later to take the world of aviation into a new era.

Two engines from the Horten 229 V3.

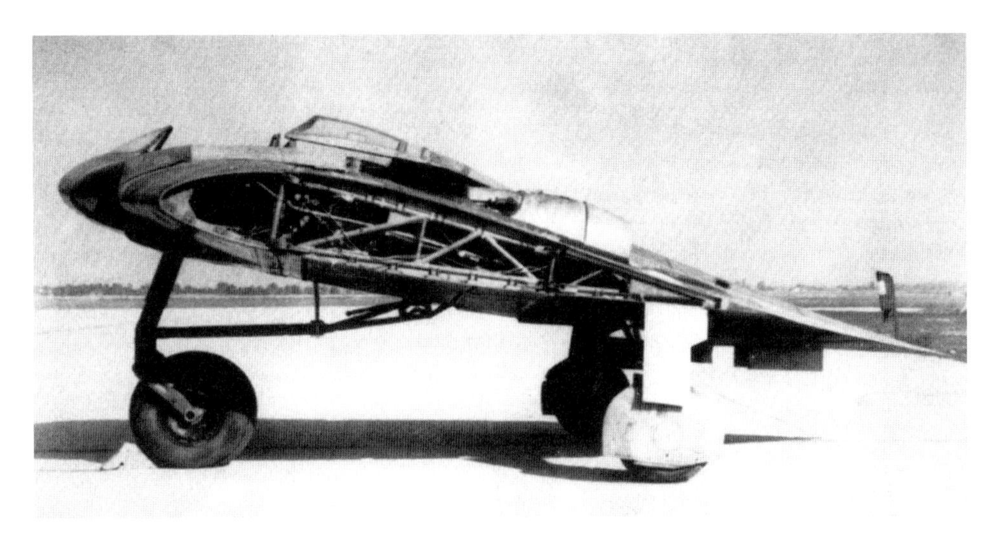

Horten 229 V3 with the wings removed showing the skeletal construction of the fuselage.

Messerschmitt P1101. The second generation of German jet fighters began with the 'emergency fighter competition' of July 15 1944 as proposed to manufacturers by the RLM. The Me P1101 VI was about 80% complete when the secret Oberammergau complex in Bavaria was discovered by Amercan troops on April 29 1945, days before the war's end.

Gerhard Fieseler's Delta IV. The aircraft crashed on its first test flight and the project was abandoned.

Cockpit and instrument panel of Me 163B.

Adolf Galland with Werner von Braun and other Luftwaffe Officers.

hired an assistant, Rudolf Nebel, who was an engineer but who had no experience with rockets. Together they worked on a number of ideas for a rocket, but nothing came of it and the film premiere came and went. Somewhat disillusioned, Oberth returned to Rumania, whilst Nebel, who had become a member of VfR, worked on a design to test liquid propulsion systems.

It was Werner von Braun who initiated the idea of using a rocket as the main propulsion unit for an aircraft and early trials were carried out, the majority of which were not successful. Despite this, experiments continued, culminating in the relative success of the Messerschmitt Me 163B rocket plane.

The first thoughts of using jet propulsion in aircraft can be traced back to the 1920s, when Italian Ingegner Secondo Campini began experiments. He had started seriously studying jet propulsion in 1929 and had built an experimental jet boat for the Italian Air Ministry in 1932. After further experiments the Air Ministry, impressed with what they had seen, gave instructions for Caproni to construct a jet aircraft and two examples of the Campini Caproni – CC.2, MM.487 and 488 – were built between 1933 and 1940. The Campini Caproni CC.2 had a wingspan of 48ft., a fuselage length of 41ft. 6in. and an all-up weight of 9,790lb. It had a top speed of 223 mph and an operating ceiling of 13,000 feet. Powered by a 900-hp Isotta-Fraschini radial engine, the first flight of MM.487, and the world's second jet-powered flight, took place on 27 August 1940 with test pilot Mario de Bernardi at the controls. Ingegner Giovanni Pedace accompanied him on the flight and the aircraft was flown from Taliedo to Guidonia for further testing. Unfortunately, it was destroyed during an air raid in 1943. What was left of the aircraft was removed and taken to Britain for examination by the Royal

Arado Ar66 trainer, later used for testing the early rocket engines that were mounted under the fuselage.

Walter Nowotny (7 December 1920–8 November 1944). Three of the Austrian-born ace's credited 258 victories were achieved in the Messerschmitt Me 262. See page 168.

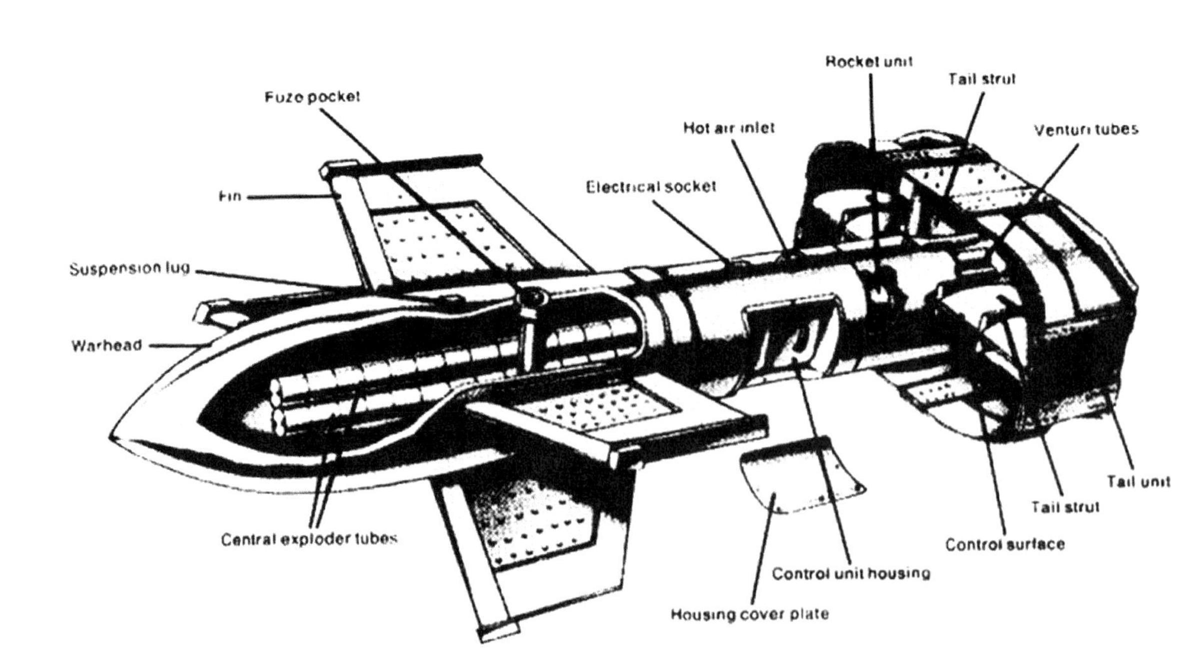

Cutaway drawing of the Fritz X bomb.

Fritz X bomb being dropped from a Heinkel He 177 during tests.

Me 163B caught in the gun camera of a P.51 Mustang.

Fritz X Bomb. The radio-guided weapon sank the Italian battleship *Roma* en route to Malta, preventing the flagship falling into Allied hands.

Air Force. The remaining CC.2, MM.488, first flew on 30 April 1941, but production of the aircraft was never considered because of the cost and the pressing military needs of the time. Fortunately the aircraft survived the war and is now on display at the Italian Air Force Museum at Vigna di Valle.

On 27 August 1939 the first jet-powered aircraft took to the air, the Heinkel He 178. The aircraft was powered by a He S-3b turbo-jet engine, which pushed out 1100 pounds of thrust, and had been developed by Hans-Joachim Pabst von Ohain whilst working for Ernst Heinkel. This led to the development of the Jumo engine when the Heinkel Company took over the secret engine projects of the recently amalgamated Bramo/BMW Company. The company had been established in 1936 at Magdeburg by Herbert Wagner and had been taken over by the Junkers Engine Division Company in 1938.

In the meantime, Willi Messerschmitt had been looking at the development of the jet engine too, and together with the RLM – *Reichsluftfahrtministerium* (German Air Ministry) conceived an engine with a thrust of 1,350lb and installed it in an aircraft with the designation P.1065. This aircraft was to be developed into the Messerschmitt Me 262, which was to take the aviation world, during the war, by storm.

The Messerschmitt Me 163B Komet rocket fighter also appeared at this time and it too made a lasting impression on the bomber crews that came under attack from the minuscule fighter.

The Junkers and Arado companies were meanwhile making progress with this revolutionary new type of aircraft.

It is these aircraft that this book considers, together with the part played by the German scientists and engineers who developed them. These rocket and jet aircraft came too late to make any significant difference to the outcome of the war, but they were to lead the way for future designers and manufacturers.

1

MESSERSCHMITT ME 163B 'KOMET'

Conception

The first thoughts on the possible use of rockets in aviation appear at the beginning of the 1900s when the noted Russian scientist Konstantin Tsiolkovsky, widely accepted as the 'Father of Rocketry', published a scientific paper in 1903 on his theory for a rocket-powered motor. Although only a theory, in his paper he specified the use of liquid oxygen and liquid hydrogen as propellants. In the 1920s two Russian students, Sergei Korolyov and Valentin Glushko, carried out experiments on gliders but the plans were dropped because of the economic climate and probably because the 'old guard' in the government disapproved of anything they did not understand. These investigations were of course reborn after the Second World War when the Russians were able to take advantage of the wealth of material acquired from the defeated German Army.

In Germany, experiments had been carried out in 1928 on liquid fuels by Fritz von Opel, the automobile manufacturer. Max Valier of the *Verein für Raumschiffahrt* – VFR Society – had also carried out such experiments with an eye to space navigation. The extremely wealthy Von Opel's only purpose for these experiments was not to further the cause of science, but to use the fuel in his automobile business and enjoy the publicity that would go with it.

Von Opel and Max Valier went to see Alexander Lippisch at the *Forschungsinstitut, Rhön-Rossiten Gesellschaft* – a research institute for aerodynamics – with the intention of persuading him to loan one of his gliders for an experiment. Lippisch had designed and built two tailless gliders, the *Storch* (Stork) and the *Ente* (Duck) and it was the Ente that was to be the first full-scale aircraft to be fitted with rockets. Alexander Lippisch had in 1921 set up a company – Weltensegler GmbH – together with his friend and fellow designer Gottlob 'Espe' Espenlaub. It was here that they designed the first of the tailless Storch gliders, which was soon followed by the building of the Storch I and the Ente.

The Ente wasn't exactly tailless; its tail was in front of the glider, much like that of the Wright Flyer. Lippisch agreed to the use of one of his gliders for the experiment and, with the help of a pyrotechnician by the name of Alexander Sander, mounted two 44lb thrust

Artist's impression of the DFS 194V-1 in flight.

Prototype Messerschmitt Me 163A V4.

Sander rockets in a nacelle fitted at the rear of the aircraft. Originally von Opel had wanted the team to mount 796lb of rockets in the rear nacelle, but he was talked out of it. The flight took place on 11 June 1928 flown by Fritz Stamer, an instructor from the Martens-Fliegerschule. The glider was launched by means of a rubber shock cord and lasted just 35 seconds. A second flight some days later, with 55lb rockets fitted, lasted 70 seconds and Stamer took the rocket-powered glider to a height of 4,000 feet and returned to the take-off point. A third test flight ended disastrously when one of the rockets exploded on take-off and destroyed part of the glider.

At the beginning of 1929, one of the students from the Martens-Fliegerschule by the name of Hatry designed a rocket-powered glider and persuaded Lippisch to help him build it. With the glider completed, Fritz von Opel viewed it and offered to buy it from Hatry. The glider was then overhauled by a workshop at Frankfurt airport to make it airworthy, and christened the Opel-Sander Rak 1 (Hatry was never paid for the machine). Whilst it was at Frankfurt, Valier and von Opel persuaded Espenlaub to modify his EA 1 glider to take Sander rockets. This he did and the glider was renamed the Espenlaub-Valier Rak 3, the first tests being carried out at Düsseldorf. In the meantime Espenlaub had commissioned Swiss engineer A. Sohldenhoff to build him a tailless aircraft that was to be rocket-powered.

The Sander rockets were fitted to the Rak 3 in a staggered manner above the wing and behind the cockpit, and a sheet metal plate was fitted to protect the vertical tail surfaces. The first two flights were uneventful, but on the third flight the tail surfaces caught fire. That was enough for Espenlaub and he decided to wait until his own aircraft, the E 15 as it was known, was ready.

Prototype Me 163A on its launching trolley.

One of the prototypes Me 163B VA+SB taking off.

The Rak 1 in the meantime had not been forgotten and Hatry continued to work and modify the aircraft. What resulted was a strengthened high-wing monoplane with a short fuselage that contained the pilot's cockpit and the nacelle for the battery of 16 rockets. The tail was mounted on two booms that were raised so as to allow the blast from the rockets to pass between and beneath. The first flight took place on 30 September 1929 with Fritz von Opel at the controls and the 16 rockets, with a thrust of 900lb, blasted him off the launching rail. Von Opel lifted the aircraft to a height of 5,000 feet and attained a speed of 95 mph. Excited by the results, he carried out a second flight some days later, but on the third flight it crashed. Fortunately von Opel was only slightly injured, but he had lost faith in the project and withdrew his support, having achieved the publicity he wanted.

This left Espenlaub and his E 13 to carry on the research into rocket-powered flight. Rockets were mounted on the trailing edge of the wing centre section and the tricycle landing gear reversed. The first tests flights however were made with a 20-hp Daimler engine. It wasn't until 4 May 1930 at Bremerhaven that the first trials were carried out using rockets, but they were not successful. Further flights were carried out with additional rockets but they too failed and when the aircraft nose-dived into the ground on the last of the flights, Espenlaube's interest took a nose-dive as well. A few days after the crash, Valier, who had been working on his own liquid fuel rocket, was killed when it exploded during a test at the Heylandt-Werke, Berlin.

At first it was thought that all interest in rocketry had gone, but unknown to anyone the *Heereswaffenamt* (Army Weapon's Department) had been investigating the possibility of using rockets in warfare and had sent a report to the *Reichswehr*. They authorised the go-ahead for research into using rockets in propulsion systems but it had to be carried out in secret. Initial research was into creating a rocket engine for possible use with missiles, but nothing significant was discovered.

The man placed in charge of the research facilities at Kummersdorf was Hauptmann Walter Dornberger and amongst his staff was a scientist by the name of Werner von Braun, – the man who, ultimately, put a man on the moon. Von Braun had been working with Valier on rocket engines whilst at the *Raketenflugplatz* together with another engineer Walter Riedel. This small team set to work designing and then developing a liquid-oxygen powered rocket of 650lb thrust. The research facilities at Kummersdorf were working with the *Germania-Werft*, Kiel, which was under the control of Hellmuth Walter.

In 1935 Walter developed a fuel based on hydrogen peroxide, methyl alcohol and liquid oxygen, which was intended to propel a torpedo that would leave no wake trace. Impressed with the results, the *Deutches Versuchsanstalt für Luftfahrt* – DVL (German Aviation

The landing skid of a Me 163B shown here extended.

Messerschmitt Me 163B with rockets mounted beneath the wings.

Me 163B being fuelled up with C-stoff ready for a test flight.

Experimental Establishment) – approached Hellmuth Walter to develop a rocket motor with a 90lb thrust for roll calibration trials.

So successful were the tests that it was decided to see what would happen if the rocket was used to assist in the take-off of an aircraft. A 300lb thrust Walter rocket was attached to the underside of a Heinkel He 72 Kadett biplane trainer fuselage. The results were better than anyone anticipated and suddenly a new and exciting chapter in the world of aviation opened. Later, further tests were carried out using a more powerful rocket mounted beneath the fuselage of a Focke-Wulf Fw 56 'Stösser', followed some months later by trials at Altenwald using the Heinkel He 111 and 116. The results were enough to persuade the RLM that the use of rockets for assisting aircraft, especially heavy bombers, was a viable proposition. They also concluded that more research into the use of the rocket motor as an engine should be explored.

In the meantime Helmuth Walter had been carrying out his own research into installing one of his rocket engines into an aircraft and his work had attracted the attention of the RLM's *Entwicklungsamt* (Development Office). So impressed were they with his development that the RLM created a *Sondertriebwerke* (Special Propulsion Systems) department to watch over his efforts. Not only did this give Walter security, it also helped considerably with the funding of his projects and it was because of this that he was able to develop an engine that operated at low temperature. This engine, known as a 'cold' engine, used a catalyst called Z-Stoff, an aqueous solution of calcium permanganate or sodium, which was forced into the combustion chamber by a pump attached to a steam turbine. This in effect enabled the two rocket fuels, when forced into the chamber, to increase the temperature to 600°C without igniting, through a chemical reaction. The engine was known as the TP-1.

Heini Dittmar taking off on a test flight in the Me 163B.

The engine was installed in a Heinkel He 112 and tests were carried out at Kummersdorf in November 1937. The tests were a complete success and although Helmuth Walter got the nod of approval from the Luftwaffe, that is all he got. Walter decided to progress further and a Heinkel aircraft was built around the rocket engine. It was called the Heinkel He 176.

Ernst Heinkel, Sigfried Gunther, Heinrich Hertel and Helmuth Walter designed the aircraft. It had a wingspan of 16ft. 5in., a wing area of 58.12 sq. ft. and was 17ft. 1in. long. Unusually it had an open cockpit and a fixed undercarriage Two prototypes were proposed, one for low speed handling tests and the other for high-speed flight trials.

In September 1938 the first of the prototypes was completed at the Heinkel facility at Marienehe. It was sent to the Luftwaffe's new, secret rocket testing site on the island of Usedom, Peenemünde-West (Peenemünde-East was the area being used by the Army). On arrival the Walter TP-2 rocket engine was installed in the aircraft. This was a moment of concern for the team, as the first Walter TP-2 engine blew up on the test stand, destroying not only the engine but also the building that housed it. A number of ground tests were carried out before the first attempt to take the aircraft into the air. At the beginning of May 1939 the first flight tests were attempted and were a complete flop. The short wings were totally ineffective and couldn't even lift the aircraft off the ground. Even with a re-designed wing and a longer runway the aircraft only managed to struggle into the air for a flight lasting 50 seconds and reached a top speed of 170 mph. A second flight a few days later in front of Hermann Göring's deputy, Erhard Milch and the Luftwaffe's head of Aircraft Procurement Ernst Udet, fared no better. Despite the recriminations they received from the two senior Luftwaffe officers (one comment from Ernst Udet was that it was a 'rocket with a running board'), the Heinkel team were delighted with the results. On 3 July 1939 the team gave a display of the aircraft to Adolf Hitler. Flown by test pilot Erich Warsitz, he climbed the aircraft to 2,500 feet, cut the engine as he swept down, then re-lit the rocket engine as he levelled out at 500 feet and completed another half-circuit before touching down in front of the Führer. Hitler was delighted and full of admiration – not for the aircraft, but for the courage and skill of the pilot – and insisted on meeting him.

The Walter rocket engine had proved itself, but the Heinkel He 176 had not.

Whilst the development of the Walter rocket engine and its trials with the He 176 had been going on, Alexander Lippisch had been designing and developing a Delta-shaped tailless aircraft. This tailless aircraft was believed by many at the RLM to be the future of aircraft design. A number of his designs, the Storch 111, IV and V, had a variety of different engines attached and were successfully tested. Using the experience and results gained, Lippisch designed a new glider, the Delta I. After initial tests, a Bristol Cherub 30-hp engine was installed and on 1 May 1931 made its first flight.

A number of other aircraft manufacturers including Messerschmitt and Fieseler became interested in the Delta design and in 1932 both these manufacturers produced a Lippisch-designed Delta wing aircraft, the Delta III and Delta IV (Fieseler F3 Wespe). Alexander Lippisch set to work to design a Delta-wing aircraft that could be fitted with a rocket engine and with the help of the *Deutsches Forschungsinstitut für Segeflug* (DFS) produced the DFS 39 Delta IVc aircraft, which was fitted with a 75-hp Pobjoy radial tractor engine. The idea of the tractor engine was to test the airframe before fitting the rocket motor, which it did successfully. In October 1937 an order came from the *RLM* for a second DFS 39 to be built with the express purpose of mating it to the newly developed Walter HWK (Hellmuth Walter Kiel) R II-203 rocket engine.

Secrecy was paramount and it was soon realised that the tiny workshop at the DFS facility was unsuitable, not only for aircraft production but also for security. It was decided that the project needed the facilities that only an aircraft manufacturing plant could provide and so Lippisch approached Willy Messerschmitt. With the approval of the RLM, the whole project (codenamed *Projekt X*) including Lippisch and 12 of his engineers and designers, was moved from Darmstadt to the Messerschmitt factory in Augsburg under the tightest security. The team took with them both the DFS 39 and their newly developed DFS 194 aircraft. Although the team operated in the confines of *Abteilung L* (Section L) in the Messerschmitt factory, they came under the authority of the Technischen Amtes of the RLM. Their first directive was that the first aircraft to be developed by the team would be known as the Messerschmitt Me 163.

Inception

Then in September the Second World War erupted and suddenly all projects came under scrutiny. It was announced from Berlin that any project that could not be completed within a year was to be abandoned. This of course put paid to the Me 163 project. The plans were shelved but Alexander Lippisch decided that unless he and his team continued with research they would be closed down. It was decided to take the DFS 194 and convert it to take the Walter R I-203 rocket engine and this was done early in the New Year. The aircraft and its new engine were taken to Peenemunde-West for testing. They had considered using the airfield at Augsburg but it was thought to be too short. By August 1940 a flight test programme by test pilot Heini Dittmar was exceeding all expectations. A top speed of 341 mph was attained, although there were some problems with excess play in the control cables that caused some difficulty in controlling the aircraft at high speeds. Placing the cables into tubes later solved this. Further tests followed and all were successful and interest in the little rocket-powered aircraft was suddenly revitalised. The *Reichsforschungsführung* (Reich Research and Planning) recommended that the programme be given a priority status.

The *Reichsluftfahrtministerium* – RLM (German Air Ministry) – ordered that work on three development prototypes be started almost immediately and they were designated Me 163 V1, V2 and V3. The engines were to be Walter HWK R II-203s and there were certain specifications laid down. The engines had to be easily removed from the Me 163 airframe for general maintenance and so that engine test runs could be carried out as and when required. The engine also had to have the capability of being re-lit in mid-air in the event of a flameout.

By October 1940 the first of the prototypes, the Me 163 V1 (KE+SW) was ready but the engines were not. It was decided to carry out glide tests with the aircraft, so it was taken to a nearby Luftwaffe airfield at Lechfeld where it was towed behind a Messerschmitt Bf 110C.

What the prototypes and planning would eventually produce. Leutnant Fritz Kelb of Jagdegeschwader 400 about to climb into his Me 163B for an operational sortie.

Tail section of a Me 163B being removed showing the small rocket engine.

Close-up of Me 163B Komet being re-fuelled with C-stoff.

The landing was carried out at Messerschmitt's airfield at Augsburg. It was staggering, that although only a glider at this moment in time, when put into a steep dive the aircraft was capable of reaching speeds of up to 528 mph. On one visit by *Generalluftzeugmeister* (Director General of Luftwaffe Equipment) Ernst Udet, accompanied by Alexander Lippisch, watched the tiny aircraft being towed up by the Bf 110C. On reaching a height of 16,000 feet, Heini Dittmar cast off and went into a steep dive. When Udet asked what the tiny aircraft was, he was told that it was the Me 163. He watched as the aircraft plummeted towards the ground and then pulled up, flashing past the two observers at a speed in excess of 400 mph with just a whistling sound. An astonished Udet turned toward Lippisch and asked what kind of engine was being used. 'It doesn't have one,' said Lippisch and when the aircraft finally landed after completing a few more passes to use up the excess speed, Udet ran across to it muttering, 'No engine. Impossible!' On reaching the aircraft he exclaimed, 'It's true, there is no engine.'

Deeply impressed, Udet went back to Berlin extolling the virtues of the new aircraft and pushing for a higher priority and more financial backing for the project. The result was that the RLM ordered a further two prototypes, Me 163 V4 and V5. Trials on the new Walter rocket engine were painfully slow and did not begin until July 1941. This of course delayed the programme, but on 13 August 1941, with test pilot Heini Dittmar at the controls, the first rocket-powered flight of the Messerschmitt Me 163 V1 took off. After one circuit of the airfield the tiny aircraft landed its limited fuel supply exhausted. One of the observers at this momentous moment was Ernst Udet and it was he that recognised the need for another test pilot. When Lippisch was asked about this he immediately requested that Rudolf Opitz be assigned to the programme. Opitz was no stranger to the Lippisch test team and had been drafted into the Luftwaffe in September 1939 as a *Flieger* (Airman). He had become involved whilst with the Luftwaffe in training pilots in the handling of the DFS 230 assault glider and had taken charge of the Hildesheim Glider School. He had just been promoted to *Leutnant* when Udet offered him the post of test pilot with the Lippisch team. Rudolf Opitz jumped at the opportunity of re-joining his old friends and getting his teeth into some real flying.

Heini Dittmar briefed Rudolf Opitz thoroughly and took him through the cockpit check and sequence. After watching Dittmar carry out a test flight, Opitz took control of the little aircraft – but he was not ready for the tremendous acceleration that awaited him as he opened the throttle. He was at 100 feet before he realised what was happening and he had not dropped the take-off 'dolly'. The dolly was a small narrow-tracked, two-wheeled, unsprung trolley that the aircraft rested on for take-off and which should be released by the pilot as soon as the aircraft lifted into the air. All take-offs had to be directly into the wind and the ground ahead had to be scrupulously examined for any uneven patches that may cause the aircraft to bounce. Landing the aircraft was carried out by lowering a centrally mounted skid from below the fuselage. Opitz thought no more of the problem and continued with the flight touching down on the grass with the 'dolly' still in position. The onlookers watched the whole episode with complete disbelief, but were relieved that the aircraft had not crashed on touchdown, as the 'dolly' had no hydraulics, nor did the pilot have any directional control over it. A testament to the pilot's skill – or luck.

Prototype Me 163B Komet taking off. This shot was taken from a film.

Moving the Me 163 around before a flight was by a *Scheuschlepper* (Shy Tug), a small tractor that was devised to tow the aircraft to its take-off point. To recover the aircraft after a flight, the same tractor had a small trailer attached that had a set of hydraulic arms that could lift the aircraft off the ground and take it back to the hangar.

The arrival of another Walter HWK R II 203A engine and its installation into the He 163 V3 (CD+IM), brought a second aircraft into the testing arena. Three flights later it was decided that it was time to attempt to take the aircraft to the 1000 km/hr. The intention was to tow the aircraft to a height of 13,000 feet and then fire the rocket engine but because the Bf. 110C could not tow a fully fuelled Me 163, it was decided to only put 75 per cent fuel load into the Me 163 V3. On 2 October 1941, Rudolf Opitz flying the Bf 110C towed Heini Dittmar to a height of 13,000 feet and released him. Dittmar fired up the rocket engine and opened up the throttle and within no time watched the airspeed indicator reach 910 km/hr (565 mph), then pass through the 1000 km/hr barrier (623 mph). Suddenly the nose of the aircraft pitched down as the tiny rocket-powered aircraft reached the sound barrier, Dittmar reduced the power, regained control and brought the aircraft down to a safe landing.

The results of the flight were kept top secret, but the Luftwaffe was fired with enthusiasm and suggestions for fitting out the aircraft with guns and rockets abounded. Using the aircraft as a fighter was one of the theories put forward but it was pointed out that the limited range and amount of fuel carried made this impracticable and its use as an interceptor was the only one feasible. Despite this, Ernst Udet wanted the aircraft put into mass production and it took a lot of persuasion by Lippisch to persuade him that the He 163 in its present form was far from being an operational aircraft. Lippisch then put forward the suggestion that he design another high-speed research aircraft. This offer was met with a deafening silence.

On 22 October, after much pressure from Udet, the Messerschmitt Company produced details for the modification and construction of 70 Me 163B interceptors. Authorisation for the project followed immediately and with the highest priority. Alexander Lippisch and his team set to work re-designing the Me 163 to enable it to carry more fuel and for

Ground crew examining the prototype Me 163B prior to a test flight.

Prototype Me 163b coming into land after its first test flight.

Pilots of Jagdgeschwader 400 admiring their new Me 163B Komet fighters.

it to be armed with machine guns. In addition Hellmuth Walter was tasked to produce a new, more powerful engine of over 3,000lb thrust for the aircraft; this was to be the Walter HWK R II-211.

The fuel for the engine was C-Stoff, which was made up of 57% Methyl Alcohol, 30% Hydrazine Hydrate and 13% water, 17% aqueous solution of Cupracyanide was added as a stabiliser. This replaced T-Stoff, which consisted of 80% concentrated hydrogen peroxide, the remaining 20% consisted of stabilisers. The T-Stoff also acted as a carrier for the oxygen that combusted the methyl alcohol and hydrazine hydrate in the C-Stoff. The cupracyanide in the C-Stoff was used to catalyse the T-Stoff and to aid the combustion of the two fuels in the combustion chamber. The C-Stoff fuel was stored in four tanks, two in each wing, whilst the T-Stoff was stored in one main tank in the fuselage and two either side of the pilot's seat. Centrifugal pumps that were linked drove both systems to the turbine pump. This in turn was driven by steam created by the T-Stoff being passed over porous stones containing calcium permanganate and potassium chromate. Being circulated around the walls of the combustion chamber before being forced through atomising nozzles cooled the resulting fuel mixture. The exploding gases resulting from all this were vented through a tail pipe and, unlike the Z-Stoff, the C-Stoff produced no tell-tale dense white smoke trail across the sky.

On 17 November 1941 as the production programme started to get under way, one of its most ardent supporters, *Generalfeldmarschall* Ernst Udet committed suicide. His

Prototype Me 163B being examined by ground crew after its first test flight.

General Adolf Galland and Major Gordon Gollob discussing the Me 163B Komet.

successor, *Generalfeldmarschall* Erhard Milch did not share Udet's enthusiasm and put the programme on a low priority footing. Production continued but not with the same sense of urgency. On the same day as Udet shot himself, the Luftwaffe sent their *Typenbegleiter* (Project Officer) to the Messerschmitt factory to oversee the development and production of the aircraft. Hauptmann Wolfgang Späte had been with II/Jagdgeschwader 54 on the Russian Front and had welcomed the re-assignment. On his arrival he was introduced to the Me 163 by Heini Dittmar and shown the controls and the cockpit layout. A couple of days later he was to fly the aircraft for the first time and, but for his skill, it could well have been his last.

After being well briefed Späte climbed into the cockpit and started the engine. As the speed of the aircraft increased it hit a bump in the ground and leapt into the air to a height of around 30 feet. Späte managed to bring the nose up as the aircraft began to settle back down and then suddenly there was another bump and the aircraft was airborne again. This time with the power full on, the aircraft started to climb, Späte released the 'dolly' and climbed away to an uneventful flight. What this highlighted was the need for the 'dolly' to have hydraulics fitted to absorb any bumps and not make the aircraft prematurely airborne. Because the rudder and elevons required a certain amount of speed before they became effective, there was a tendency for the aircraft to ground loop on take-off if it was not achieved.

The prototype Me 163B V1 appeared just four months later awaiting the installation of the Walter HWK R II-211 rocket motor, but there had been problems and it was still being bench-tested. The chances of there being an operational interceptor *Gruppe* by

Me 163B touching down after a test flight.

Operational Me 163B Komet taking off on its launching trolley.

the end of the year were diminishing rapidly. Flight tests were carried out using the Bf. 110C as a towing tug and, like its predecessors, the Me 163B V1 handled superbly as a glider.

Another problem arose at this time, the use of parachutes by the pilot. The parachute had been designed for aircrew where there was a speed of less than 300 mph, but now there was a situation where rocket aircraft were travelling at speeds in excess of 600 mph. To try and solve the problem a drogue chute was designed and fitted in an effort to slow the aircraft down to a safe bale-out speed. In the initial test with Rudolf Opitz at the controls, he was towed to 15,000 feet and released. Going into a dive and reaching a speed well in excess of a normal bale-out speed, he released the drogue chute. The chute opened and the speed of the aircraft fell away to a safe bale-out speed, but when he came to release the drogue chute the explosive charge failed. Opitz was left with a problem, he could either bale-out and take his chances that he wouldn't become entangled in the drogue chute which was still billowing out, or he could ride the aircraft down and find out why it did not release. Putting the aircraft into a steep dive in order to attain the critical landing speed he managed to land the aircraft safely in a beet field some yards from the airfield.

Wolfgang Späte, the Luftwaffe's representative, recognised the fact that the Me 163B 'Komet', as it was now called, was going to need pilots to fly it. So in between visiting the Messerschmitt manufacturing plants in Augsburg and Regensburg where the Me 163Bs were being produced, he began a training programme for pilots who would eventually form the nucleus of the Komet fighter groups. Using his personal knowledge of the most experienced pilots, Späte requested that *Oberleutnants* Joschi Pöhs, Johannes Kiel and Herbert Langer be assigned for training. As the training group got under way Späte asked for Hauptmann Toni Thaler to be assigned as adjutant to the training section. It is interesting to note that all these pilots had been serving on the Russian Front and Pöhs had served with Späte in JG 54.

The training of the new pilots progressed well and without incident, and aware of the lack of interest being shown by the Luftwaffe hierarchy in the use of the rocket-powered interceptor, Hauptmann Späte decided that the time was right to show exactly what the aircraft could do in the right hands. A display was organised at Peenemunde for June 13, 1941 in front of *Generalfeldmarschall* Erhard Milch, Albert Speer, the War Production Minister and *Generalmajor* Adolf Galland, General of Fighters. Hauptmann Späte as *Kettenführer* (Flight Leader) with *Leutnant* Opitz and *Oberleutnant* Pöhs as his wingmen, made a formation take-off and climbed in tight formation into the sky. After a brief formation flying display the aircraft landed back on the field to the warm appreciation of the guests. So impressed were they that the status of the project was raised considerably and interest reawakened.

It was around this time that a new pilot arrived for training – Hanna Reitsch, Germany's most famous aviatrix. An outstanding glider pilot, Hanna Reitsch besides being an aviator was also a celebrity, with close friends in high places. Those friends included Hermann Göring and Adolf Hitler. Willy Messerschmitt was quick to realise that to have her on his side would be a great advantage, and made plans accordingly.

By the end of 1941 the flight test and training programme was well established and Hanna Reitsch was offered the position of test pilot to the O-series of Me 163Bs whilst still continuing with her position of test pilot with the DFS in Berlin. Because all the aircraft at this time were engineless all the flight tests were of course carried out as gliders, a role which suited Hanna Reitsch down to the ground. Then tragedy struck. On her first test flight she misjudged the landing and crashed. Hanna Reitsch suffered severe head injuries that were to keep her in hospital for the next six months. Her towing pilot for the flight was fellow Me 163 test pilot Rudolf Opitz and in his opinion the injuries could have been avoided if she had followed procedures. His report stated:

> The plans on that day called for Hanna to perform the company flight test. Provided the flight of flights met company requirements, the aircraft would be turned over to me for an acceptance flight and become government property if it met contractual standards. Since the Obertraubling plant was not yet prepared for airplane tow operations, I had agreed to tow with the Bf 110 from our Augsburg operations. Eli, Lippisch's Chief of Flight Line Activities, was my observer in the back seat of the 110. After take-off we noticed that the dolly had not jettisoned, so I signalled by cycling the landing gear (lifting it up and down) before finally leaving it down. The standard procedure was to gain altitude and get acquainted with the characteristics of the aircraft, then make a decision on what to do next. The difficulty of having a job in Berlin yet remaining a test pilot now became apparent. Hanna cast off, misjudged her approach and undershot the runway. She had simply lost currency with the aircraft. The aircraft touched down in a normal landing attitude across the furrows of a freshly ploughed field. Deceleration was high in the soft soil and the Me 163 came to a full stop, undamaged, only yards away from the threshold of the runway. Perhaps nothing short of a miracle prevented the Me

> from flipping on its back during the landing on such unfavourable ground with its take-off dolly still attached to the landing skid. I am convinced to this day that Hanna would have walked away from this mishap without a scratch had she heeded the most basic of all safety precautions and worn the shoulder straps provided for the pilot and stowed away the utterly unnecessary gunsight that was mounted only inches from her face in an aircraft that carried neither guns nor an engine on its first flight.
>
> Without the protection of the shoulder straps, Hanna was thrown forward during deceleration of the aircraft and her face smashed into the gunsight, causing severe facial and head injuries.

The accident shocked her many friends and fans throughout the country as well as her critics. *Feldmarschall* Hermann Göring and Adolf Hitler sent flowers to the hospital and requested they be kept informed of her condition. Critics of the Me 163 projects saw their chance of having the programme stopped. However Opitz demonstrated the Me 163B with take-off dolly attached to the satisfaction of all concerned and was able to prove conclusively that the controllability of the aircraft was not impaired by this flight condition.

The problems did not stop there. Some weeks later at the end of November Heini Dittmar touched down in a Me 163A after a routine test flight in front of the experimental hangar at Augsburg. Those who witnessed the flight and landing saw nothing out of the ordinary when, after coming to halt, the cockpit did not open and Dittmar did not emerge. It was assumed that he was carrying out his cockpit drill before getting out. Rudolf Opitz and other test pilots strolled over to Heini Dittmar who was still strapped into his seat, and then observed that he was in considerable pain and unable to move.

Carefully they removed him from the cockpit and took him to hospital. Questioned later he told them that after landing and deceleration the aircraft had gone across deep ruts in the grass injuring his back. It was thought that the shock absorbing system in the landing skid malfunctioned and the aircraft just bottomed out, giving the pilot no protection all against the uneven ground. With Heini Dittmar hospitalised for the foreseeable future, Rudolf Opitz took over command of the Me 163 programme and brought in two more engineering test pilots, Hans Boye and Bernhard Hohmann, who were assigned to the *Erprobungsstelle* (test division) based at Peenemünde.

The delay in the producing of the Walter engine was now starting to cause problems and other engine manufacturers were approached to produce a workable engine. In the meantime tests continued on the Me 163B V1 (VD+EK) and the use of the drogue chute, and the Me 163B V2 (VD+EL) was fitted with static weapons to show the general arrangement of the equipment. The Me 163B V3 (VD+EM) was sent to the Walter Werke to be used in engine trials, whilst the Me 163B V4 (VD+EN) was used for testing different radios. The Me 163 V6 was selected for tests regarding the fitting of a pressurised cockpit. The V7 had the Mk.108 cannon fitted, but this was unsuccessful and the V8 joined the V4 in radio trials.

The friction between Willy Messerschmitt and Alexander Lippisch, which had been smouldering from day one, finally exploded and turned into virtual open warfare aggravated by the increased pressure on both of them from the Reich leadership. This was further aggravated by the closure of Abteilung L on 28 April 1943. Lippisch had had enough and left to join the *Luftfahrtforschungsanstalt Wien* (Aviation Research Establishment, Vienna), where he carried out research on the P.13 interceptor.

In May 1943 the first of the long-awaited Walter HWK 109-509 rocket engines arrived to be installed in the Me 163B V2. The pre-production Me 163B-Os had been ready at the Obertraubling facility for months, as had half of the Me 163B-1s. Along with the engine came the news that the maximum running time of the engine on full throttle was only six minutes, only half of the calculated time. By the following month the engine had been installed and once again the calculations were wrong when it was discovered that the take-off weight of the aircraft was 9,480lb as opposed to the calculated weight of 7,275lb. This was a major discrepancy as the exceeded weight was over a ton.

Despite this, the fact that the aircraft now had an engine and the feeling that things could only get better caused hope to surge through the team. Rudolf Opitz was to make the first test flight and he familiarised himself with every aspect of the engine and the aircraft. Then on 24 June 1943, he walked the whole length of the intended take-off run in the field at Peenemünde, making sure that there were no hidden ruts or bumps that would interfere with the take-off. When all was ready, Opitz climbed into the cockpit, strapped himself in and connected up his oxygen and radio leads. With a loud 'crack' the engine fired and Opitz opened the throttle. The small rocket aircraft hurtled across the field and just as he was about to lift off, the dolly wrenched itself loose dropping the aircraft on to its landing skid. Opitz, realising that he was fast running out of runway, decided to lift off and climbed into the sky. What he didn't know was that the dolly, in wrenching itself off, had damaged the fuel line and suddenly the cockpit was filled with T-Stoff (80% Hydrogen Peroxide) fumes turning everything a milky white.

Opitz struggled to find any visibility as his goggles were fogged up and the fumes were affecting his eyesight. The inside of the canopy was also clouded in a milky white vapour but then suddenly, as the aircraft's fuel ran out, it all cleared, leaving Rudolf Opitz to glide the aircraft back to the landing field. Because of the long absence of engines for the aircraft, the test pilots had had plenty of time and hours in gliding the little aircraft, so landing it without power was not a problem. This was the first official flight of the world's first rocket interceptor, the Me 163A Komet.

With the first flight came the first *Erprobungskommando* (EK) 16 unit with Hauptmann Wolfgang Späte as *Kommandeur*; *Hauptmann* Toni Thaler, Chief Technical Officer; *Oberleutnant* Joschi Pöhs, Adjutant; *Oberleutnant* Rudolf Opitz, Chief Flight Officer; *Hauptmann* Otto Böhner, Second Technical Officer. *Hauptmann* Robert Olejnik was given command of *Flugstaffel* (Flying Squadron) 1.

The EK 16 unit immediately went to work training pilots to fly the Me 163A, as it was they who were to introduce the aircraft into the Luftwaffe and by the end of the year it was hoped to have the first operational squadron in action. As well as instructors the unit brought in their own technicians to service the little rocket plane, and the man to oversee all this was *Oberleutnant* Otto Oertzen who had been one of the senior engineers at the Walter Werke.

Time was now of the essence, the engines were over a year late in coming so the pressure was on to get the engines into the Me 163Bs and the aircraft tested.

Once again there were problems when on 30 July 1943, Rudolf Opitz took a newly engined Me 163A up for a test flight with full fuel tanks. The lift-off went well, as did the climb to 26,500 feet, when suddenly violent bursts of power started coming from the engine and a fire warning light appeared in the cockpit. At 28,000 feet Opitz cut the power and glided the aircraft down to a lower altitude and attempted to restart the motor. This failed and Opitz was left with the decision to abandon the aircraft and bale out, or take it back to the field with almost a full fuel load aboard. The latter was tantamount to flying a high explosive bomb. Opitz didn't hesitate and brought the aircraft down in a steep dive before landing it back on the airfield. The hours of gliding practice in the absence of no engines for the aircraft had paid off. After hours of bench testing the engine, nothing was found that could have caused the problem and it remained a mystery and cause for concern.

Even with all the test pilot/instructors still flying the Me 163A daily, it was a steep learning curve for pupils and instructors alike. None more so than for Rudolf Opitz who on one test flight, became disoriented when climbing through a heavy haze that blended perfectly with the colour of the sea below. Managing to find a reference point through the haze (a tiny island under his left wing), Opitz suddenly realised that he was in a steep dive and curving to the left. Slowly easing the aircraft back into level flight, he levelled off at a mere 100 feet off the water and slowly regained control. On returning to the field, much to the relief of his ground crew who had last seen him plummeting toward

Me 163B dropping its launching trolley just after take off.

Me 163B about to touch down with its flaps full down.

the ground, it was found that the rudder was gone completely, and some of the fairing fasteners had pulled out due to the stress the aircraft had been subjected to during the dive.

One of the things that was of great concern to the pilots and ground crew, were the fuels – T-Stoff and C-Stoff. They had to be handled with a care that didn't allow for any mistakes, as just about any would invariably turn out to be fatal. The cockpit was made up of the fuel tanks, one either side and one right behind the pilot's seat. If the tanks ruptured they would almost immediately melt the pilot, so a flight-suit made of Asbestos-Mipolamfibre was made. The problem, it was discovered, was that the fuel could soak through the seams and the fabric itself and would ignite on contact with the skin of the wearer. The T-Stoff had the ability to eat its way through steel, iron and rubber, so it was stored in aluminium containers that were painted white, as was anything that was connected with the fuel.

Training of the new pilots began in the two-seat glider the *Kranich* (Crane) and after some hours in learning the landing techniques of the glider they progressed to the single-seat *Habich* (Hawk) that had a landing speed of around 62mph. This was a necessary part of the pilot's training, as when the Me 163A's engine was shut down it became a glider and unlike the regular type of glider, did not have to ability to pick up thermals and use them to gain height. The 'Anton', once the engine had shut down, had to be landed on the first pass. The first eight Me 163As that were built were assigned to EK 16 for training purposes. The name 'Anton' was derived from the letter A after 163,

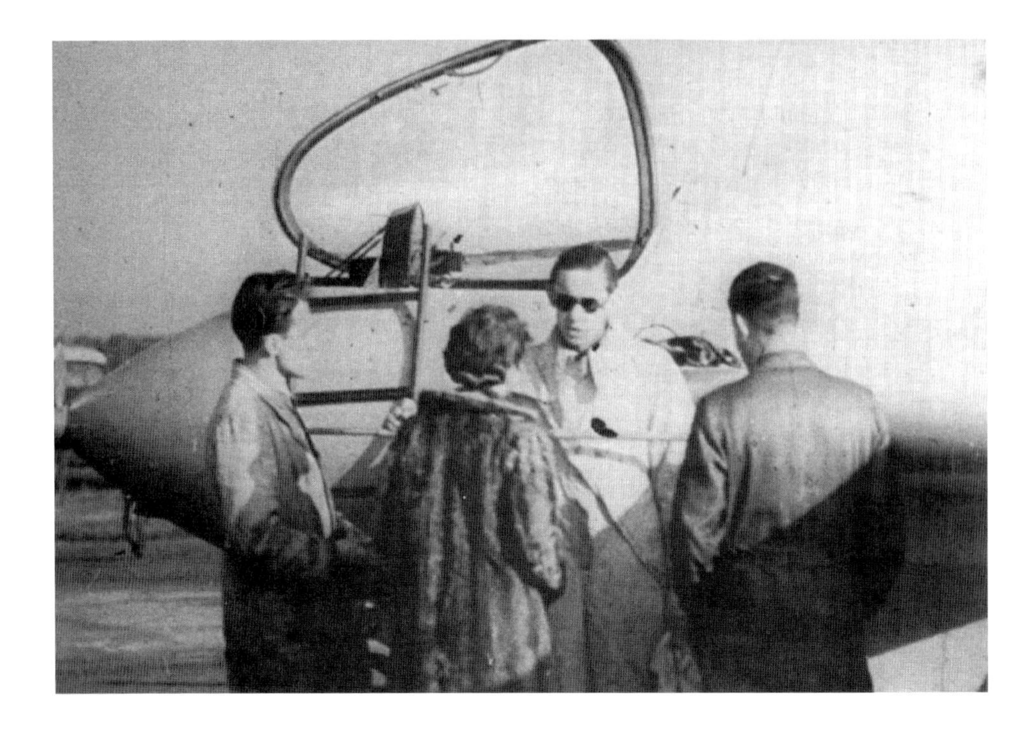

Test pilot Hannah Reitsch talking with Heini Dittmar prior to a test flight.

Me 163B taking off, the launch trolley falling to the left of the photograph.

and the B-model was known as Berta. After carrying out a number of towed starts in the aircraft and mastering the landing techniques the pupils were ready for their first 'sharp' start. A 'sharp' start was flying a fully fuelled Me 163A from a standing start. The Walter rocket engine was the HWK R II-203b 'cold' engine that, for normal take-off, would read 8,000 rpm, 20 atmospheres of pressure and an engine exhaust temperature of 600°C. The rule was that if any of these figures dropped during take-off, the pilot was to abort and cut the engine.

Because of the delay in producing the Me 163B Komet, the only person until September 1944 to actually fly a rocket powered B-model Komet was Rudolf Opitz. On the face of it both aircraft appeared to be the same; but the transition from the Me 163A to the Me 163B was enormous. The landing speed of the B-model was significantly higher because of the remaining fuel that was in the tanks. When fully laden the B-model weighed 8,367lb, the fuel consisting of 1,032lb of C-Stoff and 3,417lb T-Stoff. It was discovered during tests was that the rapid rate of climb from sea level to the rarefied atmosphere of the higher altitudes subjected the pilots to a variety of physical changes hitherto unknown to them. Fortunately, a low-pressure chamber had been acquired from the Wermacht, who had captured it in Russia, and the chamber was sent back to Germany for high-altitude research. Dr. Erich Dunker, a physician who had been assigned to EK 16, acquired the chamber and he was able to subject the trainee pilots to 'rides' of up to 40,000 feet at 'speeds' faster than that of the rocket-powered Komet. Results obtained

RAF photo-reconnaissance shot of two Me 163B aircraft on an unknown airfield.

Close-up of ground crew re-fuelling a Me 163B Komet with C-stoff. Note the protective suit of the ground crew.

Me 163B being wheeled out of its hangar.

The launching trolley of a Me 163B being manoeuvred into position.

from these tests enabled the doctors to create a special diet for the pilots that would not produce intestinal gases that would expand at high altitude causing extreme discomfort, and in some cases, serious problems for the pilots. The cockpit sequence for the pilot was as follows:

1. The pilot was strapped into the cockpit by his mechanic and the instruments were checked.
2. The control surfaces were checked by testing the controls for freedom of movement and the elevator trim was set 3° to 6° tail heavy.
3. Oxygen mask was fitted to the pilot's face and an oxygen flow check carried out.
   Mechanic closed the canopy which was locked from the inside by the pilot.
5. The throttle had five positions: Off; Idle; 1st Stage; 2nd Stage; 3rd Stage. The throttle was moved to the idle position, which exposed the starter button. This was pushed and the firing sequence began. The T-Stoff turbine was activated and this pumped the fuel into the power plant. After five seconds the pilot released the starter button with the engine firing and the tachometer reading between 40% and 50%.
6. After a further check of the instruments the throttle was moved to the 1st Stage, then

Low-level high-speed flight of a Me 163B during testing.

the 2nd Stage. Another check of the instruments and on to the 3rd Stage. At this point the engine, now on full power, would jump the small chocks holding it back and begin the take-off run.

7. At between 20 and 30 feet the dolly would be jettisoned and a steep climb initiated by the pilot.

Throughout the training and transition periods at Peenemunde, the pilots were not only under the scrutiny of the instructors and the German hierarchy, but they had been under observation by the Allies. A RAF de Havilland Mosquito on a photo-reconnaissance flight

Excellent shot of a Me 163B Komet coming into land.

Me 163B Komet taking off on its launching trolley.

on 26 July 1943 had spotted the tiny tailless aircraft on the runway. This new phenomenon concerned the Allies and confirmed their worst suspicions that something out of the ordinary was going on at Peenemunde. The simple answer was to initiate a heavy bombing raid and this was carried out on 17 August 1943 by 126 B-17 Flying Fortresses belonging to the 8th Air Force. The attack was carried out on the Messerschmitt A.G. at Regensburg

An operational Me 163B of JG400 making a 'sharp' start.

Leutnant Fritz Kelb wearing his asbestos flying suit climbing into the cockpit of his Me 163B.

destroying 11 Klemm-built Me 163B-0s and bringing the Bf 109 production line to a grinding halt.

This was to have a severe knock-on effect upon the Me 163B production line at Obertraubling as the production of the Bf 109 was moved from Prüfening and production of the Me 163 stopped. This meant that all production of the Me 163 was now centred around the tiny Klemm factory, where knowledge of building such a complex aircraft was almost nil. In addition the company had to 'farm out' work to the Dornier Oberpffenhofen and Bachmann von Blumenthal companies.

Within hours of the decisions being made, the RAF struck again as 597 heavy bombers from Bomber Command unleashed their deadly cargoes on Peenemunde. The V-weapon section at Peenemunde-East took the brunt of the attack, whilst Peenemunde-West, where the EK 16 was, suffered only minor damage. But they were now exposed and open to attack at any time so it was decided to move the programme, lock, stock and barrel to Anklam, situated on the mainland close to the tip of Usedom Island. During the removal of the aircraft, Rudolf Opitz flew one of the Me 163Bs himself because it had faulty flap hydraulics. On take-off he found that the dolly wouldn't jettison, a very similar incident to the one that Heini Dittmar experienced a year earlier, but he continued with the flight. As he approached to land he realised that the landing-skid hydraulics had ceased to function, so he gently put the aircraft down on the retracted skid. Unfortunately over the last few yards the aircraft had to run across a series of deep ruts and the jarring injured Opitz's back, badly damaging two of his vertebrae. Rudolf Opitz was hospitalised for three months, but even then he managed to keep in touch with his command, persuading someone to let him have a Fieseler Storch that he flew to and from the hospital, much to the annoyance of the doctors.

The almost identical injuries caused to both men arising out of similar circumstances prompted Opitz's physician, Dr. Schneider, to investigate how the injuries were caused. After a great deal of research he realised that cushioning the aircraft itself was impractical, but he devised a torsion sprung seat for the pilot that worked well and was put into all Me 163B Komet aircraft.

The move to Anklam was a short one as the Luftwaffe felt that their move had been spotted by one of the RAF's photo-reconnaissance Mosquitos. By September 1943 the EK 16, consisting of 5 instructor pilots, 23 student pilots and 122 ancillary staff, had moved to Bad Zwischenahn near Oldenburgh. The moves caused serious problems in the training programme because EK 16 had to share the airfield with a number of other Luftwaffe *Jastas*. Maintenance hangars were difficult to come by, as was accommodation. Training flights had to be arranged around the normal everyday operations of the squadrons using the airfields. *Hauptmann* Wolfgang Späte and *Oberleutnant* Rudolf Opitz pressed for a ground radio homer to be fitted to aid the Me 163 Komet pilots during training. The request was refused and so left the flight training dependent wholly on good weather, and of course ruled out any night flying.

Engine problems dogged the programme and one in particular was only discovered after extensive bench tests. When installed in the Me 163A the engine cut out just after lift-off but fortunately the pilot was able to get the aircraft back safely. When the engine was removed and bench tested it operated perfectly every time, until it was discovered that there was a

The specially designed Scheuschlepper being used to recover a Komet after a test flight.

faulty venting system that did not operate fully when the aircraft was at a slight incline. The continuous acceleration and the build-up of turbulence at the T-Stoff outlet caused bleed air to be sucked in with the fuel, resulting in the automatic fuel control shutting down. Installing vertical baffles above the outlet solved the problem.

On 30 December 1943 *Leutnant* Joschi Pöhs was killed when he attempted to land after an engine malfunction. His wingtip touched a flak tower as he was lining up the aircraft to land causing him to lose all control and the aircraft, fully fuelled, exploded on contact with the ground.

The New Year highlighted the rapid deterioration in the aerial defence of Germany. The long-range heavy bombers of the Allies were now being accompanied by very fast, powerfully armed fighter aircraft, P-47 Thunderbolts and P-51 Mustangs. It also brought the first three Me 163B Komets to be built by the Klemm facility to EK 16. After flight tests by Rudolf Opitz it was found that they all had serious problems and these would have to be ironed out before they could be accepted and channelled into the training programme. But tragedy struck EK 16 again when one of the most experienced and highly skilled trainee pilots, *Feldwebel* Alois Wörndl, was killed after a training sortie in a Me 163A. He had been the first of the trainees to make a 'sharp' start in the Me 163A, and the accident happened during one of his flight training sessions, when, after his fuel had been expended, he carried out a steep dive over the airfield resulting in a low pass across it. Soaring back up to go into the landing pattern, he took the aircraft outside its gliding range and consequently found he had to put down in a field close by. The aircraft skidded over the ploughed field then turned over, throwing Wörndl out of the cockpit just before it exploded. Wörndl's body was found a few yards away, his neck and both legs broken.

The three Me 163B Komets that had been delivered from the Klemm factory had not been flight tested at the factory for security reasons and because of the scarcity of the fuel. Opitz and Späte decided to continue to try and iron out the problems with the Me 163B and continued test flights. The January weather had turned bitterly cold causing heavy frosts and frozen ground. This of course created a problem whilst landing and the aircraft, because it had no brakes, skidded a further 50%. In one incident Wolfgang Späte touched down at one end of the field and then suddenly found himself skidding toward the barbed-wire fence at the far end. The aircraft rolled gently on to its back and Späte stayed until the rescue truck arrived and freed him. A second similar incident two weeks later put him in hospital with concussion after he had jumped from the aircraft as it skidded across the field. His reason for jumping was that he did not want a repeat of hanging upside down in the cockpit of the aircraft whilst what was left of the extremely volatile fuel on board leaked out around him. The aircraft had a mechanism on board for dumping the remaining fuel, but both Opitz and Späte had never activated it, but instead chose to take a measurement of what was left in an effort to find out a more exact duration and range of the Me 163B.

The training continued and by the end of February 1944 all the trainee pilots had carried out 'sharp' starts in Me 163B. The speed of training had been aided by the recruitment of three civilian engineering test pilots, Karl Voy, Franz Perschall and Otto Lamm after they had gone through training on the Me 163B at Zwischenahn. The Klemm factory was now in full production and within weeks a total of 70 Me 163Bs had been delivered to EK 16. The Messerschmitt factory was also back in production and the first 40 of their aircraft were retained by the factory to be used in a test programme they had set up in conjunction with the Luftwaffe. Their aircraft were painted in an all-grey livery and were allocated V (*Versuchs)* numbers.

Problems still continued with the Klemm-built Me 163Bs, the main one being the wing-to-fuselage fairing, which was one-piece and never seemed to fit properly. It has to be remembered that the Klemm factory was never equipped to deal with such a large order. It was to be some months before the first Me 163B Komet was considered to be ready for *Kommando* service and it was not until January 1945 that the first combat-ready Me 163B V14 (VD+EW) arrived at Zwischenahn. Rudolf Opitz was selected to carry out the initial test flight and the first simulated attack on a bomber formation. During the first simulated attack, which was set at 26,000 feet, the Me 163B was to attack from the rear. However, on reaching 20,000 feet the *Komet's* engine flamed out and it took over two minutes to carry out a mid-air restart of the rocket engine. This under combat conditions would have left the *Komet* completely vulnerable to attack from any accompanying fighters. After a number of simulated combat attacks it was realised that there was a problem that was not going to go away and the aircraft was returned to the factory to undergo tests. The fault was traced to the automatic fuel cut-off that controlled the flow mixture of the C-Stoff and T-Stoff. In the event of there being uneven amounts of fuel pumped into the combustion chamber, the engine would shut down automatically. Then as the aircraft decelerated, air became trapped in the fuel outlet creating a vapour lock that activated the cut-off system.

Whilst the scientists and engineers were sorting out the problem and rectifying it, the Luftwaffe High Command created the first Me 163B operational unit. It became known as *20/JG 1* (*20th Staffel* of *Jagdgeschwader* 1) based at Zwischenahn. Interestingly enough, at the time of its conception 20/JG 1 had neither aircraft nor pilots. One month later, under the command of *Hauptmann* Robert Olejnik, the unit was re-designated *1/ Staffel* of *Jagdgeschwader* 400 but even then only had a couple of aircraft and pilots assigned to it. The Me 163B had originally been armed with the 20mm cannon, but this was found to be ineffective against the heavy American bombers, so it was fitted with a 30mm cannon.

The development of the Me 163B had now reached the stage where it had to enter operational service, even though there were still problems with it.

The Komet Goes to War

The first operational sortie to be carried out by the Me 163B was on 13 May 1944 and was flown by *Hauptmann* Wolfgang Späte. His aircraft, Me 163B V41, took off watched on radar by Bad Zwischenahn's *Wurzburg-Riese*. After a few minutes two dots were spotted on the radar screen and the information passed on to Wolfgang Späte. As he hurtled toward the two aircraft faster than he had ever travelled, he recognised them as P-47 Thunderbolts. The two aircraft were ahead and above him and as he closed on them he cocked his machine guns and his MG 151/20 cannon. Just when he had manoeuvred into position, his engine cut-out, and suddenly he was no longer the hunter but potentially the hunted. Späte realised that he would have to wait at least two minutes before attempting a re-start and immediately started the stopwatch on his instrument panel. Trying to maintain height and distance behind the two American aircraft, he watched them disappearing into the distance. With the two minutes up Späte hit the starter button and watched the tachometer gauge respond immediately. Späte watched the turbine reach 40% then advanced the throttle, the aircraft responded with a leap forward and he raced after the two P-47 Thunderbolts.

As he closed he prepared his guns once again and watched as one of the aircraft got bigger and bigger in his gunsight. He was poised to press the firing button when the left wing suddenly heeled over and the whole aircraft began to shake violently. This was followed by the engine flaming out, at which point the aircraft's speed bled off and the shaking ceased. He had inadvertently touched the 'Sound Barrier', as it was later to be known. Putting the aircraft into a glide he returned to the airfield at Bad Zwischenahn, annoyed that he hadn't manage to open fire on the enemy aircraft – but pleased to have survived the ordeal.

Whether it was due to this incident or something else, Wolfgang Späte was suddenly transferred back to his old unit *JG* 54 to take command of the fourth *Gruppe* there. The new *Kommandoführer, Hauptmann* Toni Thaler took his place, but in overall command was *Oberst* Gordon Gollob, one of the Luftwaffe's most highly decorated fighter pilots. Why the Luftwaffe chose to move Späte is a mystery, because he and Rudolf Opitz had been responsible for creating the training programmes and taking this totally new innovation into the war. It needed a strong visionary to see where this programme could go and Oberst Gollob lacked this quality. His strengths lay with proven aircraft, not with experimental models and his lack of commitment and vision was to stifle the Me 163B programme. Toni Thaler had been with EK 16 from the beginning and so knew what was required but he was controlled by Gollob.

There was a report on 16 March 1944 by the pilot of a Mosquito, Flying Officer M.R.Hays, RAF, that he had come under attack from two Me 163Bs whilst on a photo-reconnaissance flight over Leipzig. Hays states that he was at 30,000 feet when two Me 163Bs appeared either side of him and commenced a simultaneous beam attack. He carried out evasive manoeuvres taking his aircraft down to treetop height before managing to give them the slip. During the encounter his starboard engine was hit and was pouring smoke all the way back to Lille, where he managed to land safely. This report however does not seem to coincide with the Germans report that their first operational sortie was not until May.

Whilst all this was going on the Allied photo-reconnaissance aircraft were flying over Bad Zwischenahn and were noticing the build-up of Komet fighters at the new operational base. A bombing raid was launched on 30 May 1944 and the *Kommando* lost two Komets and four were badly damaged. The main damage was to the field itself and was so bad that all flying was suspended for a number of weeks. In the meantime *Hauptmann* Thaler took the unit to Brieg to get it back into operational shape. EK 16 were not the only operational unit flying the Komet, *No.1 Staffel* at Wittmundhafen and *No.2 Staffel* at Venlo were flying sorties against the ever-increasing heavy bomber raids. The reports going back to England of tiny

Concrete bunkers used for assembling the Me 163B Komet.

Rudolf Opitz being prepared for a test flight.

aircraft travelling at incredibly high speeds showed the confusion that reigned amongst the Allies. It was not being aggravated by any German success. A lot of this was put down to the continuing engine problems that were haunting the programme. In one incident three Me 163Bs from *No.2 Staffel* attacked a formation of B-17 Flying Fortresses, but before they could commence their attack, one of the Fortresses was hit by flak and destroyed, and two of the Me 163Bs suffered engine flameouts.

Another encounter took place on 31 May 1944, this time between a photo-reconnaissance Spitfire, and a Komet from *No.1 Staffel.* The Spitfire pilot was at 37,000 feet and about to make his first run over the target, when he noticed a white trail some 7,000 feet below and about a mile away. The Spitfire pilot climbed his aircraft to 41,000 feet and saw the tiny aircraft just 3,000 feet below him and about 1,000 yards away. The sudden realisation that the tiny machine had climbed around 8,000 feet in the same time that the Spitfire had climbed 4,000 feet was quite disturbing. The only description the pilot could give of his potential adversary was that it appeared to be all-wing and travelled exceptionally fast.

Problems were still being experienced with take-offs and landings. The unreliability of the jettisoning of the dolly was still a concern and the landings on just a skid without any form of braking system were positively dangerous, especially on frozen ground. The Luftwaffe ordered an investigation into the possibility of fitting a retractable undercarriage to the rocket interceptor and a new engine that would have a longer running time. By June 1945 a new engine had emerged from Walter, the HWK 109-509C. The engine had an auxiliary chamber fitted, which increased the thrust from 3,750lb to 4,410lb and was installed in the Me 163B V6 and V18. The tests were carried out by Heini Dittmar, who had recently been released from hospital after his accident. All results were better than expected. On one flight however, Dittmar had both chambers open and had reached a speed of 702 mph when the aircraft started to pitch violently. Like Rudolf Opitz, Heini Dittmar had reached the 'sound barrier'. After struggling to bring the aircraft under control Dittmar brought it down to a safe landing. He then discovered that the fuselage was slightly bent and the rudder had been torn off.

With the results of the tests being far better than anyone could have hoped, bearing in mind the ongoing problems that had dogged the programme, it was decided to install the new engine in an improved model of the Me 163B, the 163C. Using the same wing as the Me 163B, the fuselage was enlarged to accommodate bigger fuel tanks and hold more ammunition for the MK 108 cannons, which had been relocated in the fuselage. In addition

Rudolf Opitz being assisted into the cockpit of his Me 163B Komet. Notice the cumbersome asbestos flying suit and boots.

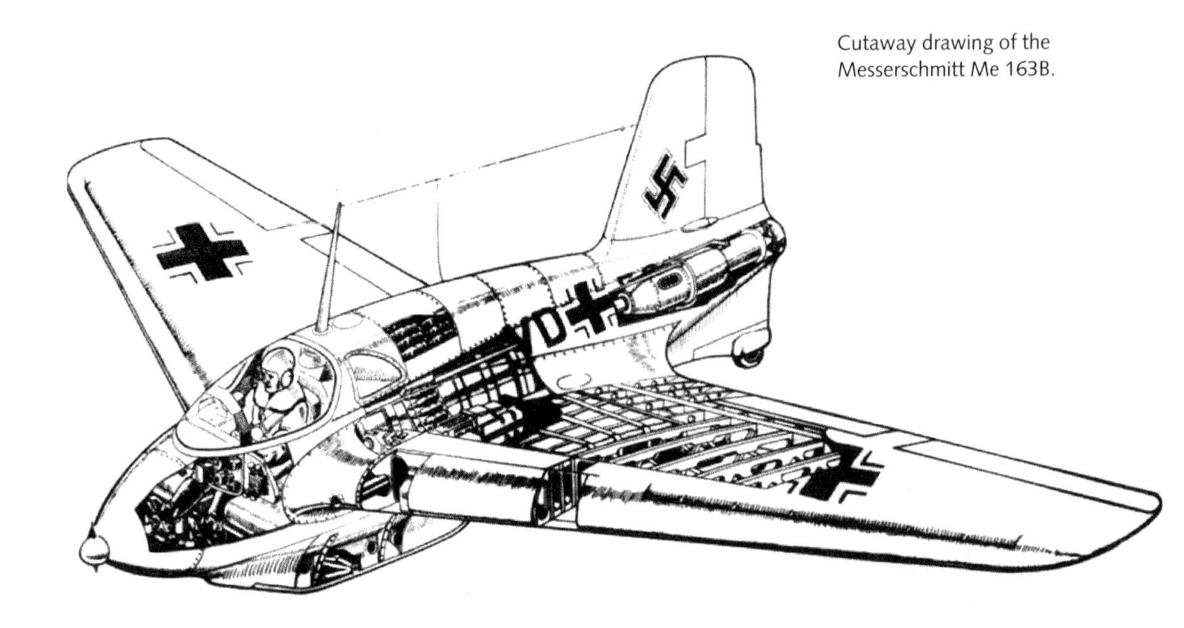

Cutaway drawing of the Messerschmitt Me 163B.

a pressurised bubble canopy was fitted, but only three models of this aircraft were built, as a third model, the Me 163D was quickly on the scene. The airframes for the D-model were manufactured by the Focke-Achgelis company at Delmenhorst but the Messerschmitt factory were now involved in so many new designs that the RLM decided to pass the project on to the Junkers company. Re-designated the Junkers Ju-248 the aircraft had a retractable tricycle undercarriage and a pressurised cockpit.

Messerschmitt continued to build the Me 163B. The only other new model they produced was the Me 163S. This was a two-seat trainer and because of the demand for the fighter version of the aircraft, only a couple were actually built and flown.

Nos. 1 and *2 Staffeln* were brought together to form *I/JG 400*, under the command of *Hauptmann* Robert Olejnik. This was not the norm, as it usually required three *Staffeln* to form a *Gruppe*, but these were unusual times. The *Gruppe* was now officially described as being *Objektschutz-Jägers* (Target Interceptors) and were assigned to protect Brandis. It was here that the Leuna Synthetic Fuel and Rubber Manufacturing plant was situated, the source for commodities desperately needed to replenish Germany's rapidly depleting stocks of war material. The arrival of these strange aircraft astonished the soldiers already stationed at the plant.

Even though the aircraft had been operational for some time and most of the problems that had dogged the aircraft had been ironed out, there were still accidents. During the next two months, three Me 163B Komets blew up on the runway leaving nothing but charred remains of pilot and aircraft all over the place.

The commanding officer *Hauptmann* Robert Olejnik was determined not to lower the morale of his men so whilst at Wittmundhafen he chose to make the first 'sharp start' from

Excellent side view shot of an operational Messerschmitt Me 163B.

Remnants of the rudder from a Me 163B after a high-speed flight.

Scheuschlepper (Shy Tug) being used to tow a Me 163B to a take-off position.

there. Everything went well after take-off and he climbed the little rocket plane and was passing through 10,000 feet when suddenly the engine started to falter and clouds of steam started to come from the exhaust. Cutting the power he put the aircraft into a glide and after two minutes attempted a re-start. As he did so, black smoke streamed from the exhaust. Dumping the remaining fuel in the tanks, Olejnik headed back toward the field, jettisoning his canopy as he approached. The aircraft touched down hard and bounced back into the air. As it came down again the port wing dipped and dug into the ground sending the aircraft cartwheeling across the field. Olejnik had obviously released his harness as he was thrown clear of the aircraft seconds before there was an almighty explosion and the aircraft disappeared into a thousand pieces. He suffered broken vertebrae, bruises and cuts, and the morale of *JG 400* was dented badly. One month later *JG 400* moved its entire operation to Brandis near Leipzig.

Just as the pilots enjoyed a certain prestige in their role as rocket pilots, so did the mechanics that maintained the aircraft. They were specialists in their own right and assumed a certain elitism whilst working on these top-secret projects.

One of the first major encounters with a large bombing force occurred on 28 July 1944, when 569 B-17 Flying Fortresses were intercepted on a mission to bomb the Meresburg-Leuna petroleum complex. The complex was under the protection of *I/JG 400* and seven of their Me 163B interceptors rose to the challenge. All the training and simulated attacks on bomber formations was about to be put to the test. The canopies slammed shut, the rocket engines were ignited and minutes later seven rocket-powered aircraft climbed into the sky.

Climbing above and behind the formation of bombers the seven interceptors raced in to attack. All seven of the pilots completely misjudged their closing speeds and shot between and below the bombers. The P-51 Mustangs turned to head off the next pass, but the Komet pilots were all experiencing problems with their engines flaming-out and the end result was that they never got near enough to get off any shots. Because of the exceptional closing speed of these little aircraft, it was soon realised that only an exceptional marksman was going to be able to fire accurately. In addition to this, the MK 108 cannon were constantly jamming, so an alternative had to be found.

The Hasag Company in Leipzig came up with a very unique automatic firing system. Named the *Jaegerfaust* (fighter fist) it comprised ten 50-mm gun barrels mounted vertically in the wing roots in a splayed, fan-shaped fashion, five on each side. Each barrel, which fired upwards, was armed with 2,2-pound high explosive shell. To counteract the recoil, a counterweight weighing the same as the shell was fired downwards to compensate for the discharge. The guns were activated by means of a photoelectric cell, which detected the shadow on the enemy aircraft as the Me 163 passed underneath. On the first real test run carried out by *Leutnant* Gustav Hachtel, the pilot switched on the photoelectric cell whilst about 1,000 feet from the ground and suddenly there was a violent explosion. Hatchel's head was slammed back against the headrest, the canopy suddenly disappeared and his helmet

Test pilot Rudolf Opitz in the cockpit preparing for a test flight.

Overhead shot of Me 163B VD+EL, one of the early prototypes.

The remnants of a Me 163B after exploding when the wingtip touched the ground whilst on take-ff causing the aircraft to flip over.

and goggles were ripped from his head. With his vision seriously impaired Hatchel managed to set the aircraft back down onto the airfield after a series of heavy bumps from which he suffered a fractured spine. It was found later that the moment he switched the photoelectric cell on, all ten shells had gone off at the same time. Modifications were carried out and twelve Me 163Bs were equipped with the *Jaegerfaust* but only one ever became operational.

29 July saw six Me 163B Komets on standby after 647 bombers from the 1st and 3rd Divisions of the 8th Air Force carried out another raid on the petroleum complex at Leuna. The six were scrambled, not as the bombers arrived but after they had completed their mission and were on their way home. Three of the bombers had been brought down by anti-aircraft fire and a number were seriously damaged. It was these that the interceptors would be after but P-38 Lightnings of the 479th Fighter Group had been called up to escort the stragglers home. After two or three runs at the stragglers the Komet pilots called it a day, mainly because they did not have the fuel capacity to carry out a sustained attack and because of the ongoing engine problems of flame-out.

The persistent problems with the Me 163B's engine usually meant that the maximum number of operational Komets at any one time was 10. This was intensely frustrating for the remaining Komet pilots who watched their comrades take off and go into action. Then the 16 August 1944 saw the first casualty of *I/JG 400*. Five of the aircraft had been scrambled after 1,096 B-17s and B-24s had come within combat range of the little interceptor. As they climbed into the sky to intercept the bombers they came under intense fire from the gunners inside the bombers. *Feldwebel* Herbert Straznicky singled out a B-17 from the 305th Bomber Group and attacked it from the rear. The rear gunner in the B-17 waited until the little rocket aircraft was within 1,000 yards and opened up a withering hail of fire. Straznicky managed to get off one burst before his aircraft was raked with machine gun fire. With smoke pouring from his aircraft, *Feldwebel* Straznicky bailed out and landed safely.

Another of the Komets flown by *Leutnant* Hartmut Ryll closed in on another of the B-17s and raked it with gunfire killing the waist gunner and causing severe damage. On a second pass he again raked the crippled bomber killing the ball turret gunner and inflicting more

Head-on view of the prototype DFS 194 on its launching trolley.

Another view of Rudolf Opitz in asbestos-flying suit climbing into the cockpit of a Me 163B Komet. No room – and very little room for error.

damage. But as he turned for the third pass and the *coup de grâce*, he himself was hit by machine gun fire from the leader of two escorting P-51 Mustangs from the 370th Fighter Squadron who had suddenly arrived on the scene. After a tussle that saw the aircraft racing all over the sky, *Leutnant* Ryall's Komet was seen to take a number of hits in the cockpit and fuselage. Flames suddenly leapt from the aircraft and it went down, the first fatality under combat conditions for *I/JG 400* and the first 'jet' aircraft to be shot down by the Allies, claimed by Lieutenant Colonel John Murphy, AAF.

The whole attack by the Me 163B Komets had lasted only 15 minutes from start to finish and the *Gruppe* claimed one B-17 Flying Fortress. It was later discovered that although it was very severely damaged, the pilot of the B-17, 2nd Lieutenant C.J.Laverdiere, AAF, managed to get the aircraft back to base in England.

In a later sortie, Leutnat Fritz Kelb became the first, and only pilot, to claim a victory using the *Jaegerfaust* when he shot down a B-17 straggler, on 10 April 1945. Fritz Kelb was himself shot down and killed one month before the war ended.

For the next ten days the mechanics worked night and day to get the maximum number of Komets ready for the bombing raids. Fortunately for them the Allies were giving other areas of Germany a pasting, leaving the area around Brandis relatively peaceful. Then on 24 August 1944, 1,300 B-17 Flying Fortresses and B-24 Liberators headed for Brüx, Leuan, Ruhland and Weimar. On the ground at Brandis eight Me 163B Komets were ready and were

Ground crew about to unlock the cockpit of Rudolf Opitz's Me 163B after a test flight.

Test pilot Rudolf Opitz being assisted by ground crew after a test flight.

One of the prototype Me 163Bs being towed back to its hangar by a Scheuschlepper after ground tests.

Hannah Reitsch talking with fellow test pilot Heini Dittmar about flying the Me 163B.

launched as the huge formation of bombers approached Leuna. Climbing to 30,000 feet the eight interceptors shut down their engines and went in to attack the low group of bombers. By shutting down their engines they left no telltale trail in the sky for the gunners to watch for. They had been seen approaching the area earlier, but such was their closing speed that the first two were through the formation and climbing away before anyone realised. On the first pass by *Leutnant* Siegfried Schubert, one of the B-17s was hit in the wing and the number four engine set on fire. As the B-17 pulled out of the formation it was hit again and was sent spinning down, exploding before it hit the ground. A second B-17 was attacked from the rear, but as the Komet streaked past, the tail gunner shattered its tail with a long burst and it was last seen heading straight down.

A Messerschmitt Me 163B being paraded around America after the war.

Schubert now had the feel for the battle and attacked another of the B-17s. On the first pass he again peppered one of the wings of the B-17, on the second he shattered the wing, setting it on fire. The large, heavy bomber heeled over as flames enveloped it and then blew up as it plunged toward the ground. The aerial fighting seemed to go on for hours, but in fact the whole battle lasted less than half an hour. The bombers lost six of their number to the Me 163 Komets of *I/JG 400*, who in turn suffered the loss of two of their aircraft. This does not take into account the bombers that had to turn back because of heavy damage to their aircraft.

The formation of *I/JG 400* had taken away most of the training pilots from *Erprobungskommando*, who were now unable to continue with the training programme. It was decided to form a new unit *Erganzungs Staffel* to be attached to *JG 400*. This would enable the new pilots to be trained under the wing of an operational unit and hopefully make the transition from fledgling pilot to operational pilot that much easier.

The new training programme was helped by a lull in the fighting in their area as the Allies targeted other areas of Germany. *Hauptmann* Robert Olejnik had been made commander of

Feldwebel Siegfried Schubert, the most successful of all the Me 163B Komet pilots with three 'kills'.

Me 163B of JG400 ready for a mission.

the training group and *Leutnant* Ziegler, a production test pilot, was posted in to help with the instructing.

More raids by Allied bombers caused serious disruption to the operational effectiveness of the *I/JG 400*, when one of the main factories at Kiel that produced C-Stoff fuel was bombed and totally destroyed. This was further aggravated when it was discovered that one of the railway marshalling yards that was heavily bombed contained a number of tanker trains containing fuel for the rocket aircraft. It was obvious that it was the oil refineries that the bombers were after and the continual destruction and damage being caused to them was instrumental in rapidly bringing Hitler's Third Reich to its knees. The bombers had to be stopped, but this was becoming increasingly difficult to do because of the lack of fuel, spares and indeed, pilots.

During September 1944 I/JG 400 were in action almost every other day but they were having problems with the radar vectoring and were being given co-ordinates that sent them into empty space behind the bomber formations. With the scarcity of fuel becoming even more of a problem this was the last thing they wanted. The Luftwaffe High Command decided to increase the size of JG 400 with the introduction of No.3 and 4 Staffeln on 24 September 1944. The *Geschwader* had a total of 19 aircraft, of which 11 were deemed to be operationally ready. On the 28 September *JG 400* Komets were in action again over Merseburg. Six Me 163Bs streaked into the sky to intercept bombers of the 100th Bomber Group, AAF, and join up with fighters from JG 300. On reaching 30,000 feet the first of the

Me 163 VI with fixed undercarriage. First flown in February 1945 by test pilot Heinz Peters.

runs were made against the bombers without firing a shot. P-51 Mustangs from the 355th and 354th Fighter Squadrons, AAF, watched as one of the Komets climbed from 9,000 feet to 36,000 feet in a matter of seconds. Chasing these little rocket aircraft was futile and the only way to catch them was to intercept them as they made their passes against the bombers. At the end of the sortie, damage was caused to some of the B-17s, but none was shot down and a number of the Komets sustained some damage. And that attrition meant that more and more of the severely damaged aircraft now had to be cannibalised as spares for the less damaged ones. By October JG 400 had two fewer aircraft than it had the month previously.

Almost daily until the end of December 1944 the Me 163B Komets were in action, but things were not going well. The American casualties were dropping whilst the Luftwaffe's were rising. The Allies were putting more and more bombers in the air accompanied by more and faster fighter escorts. The Luftwaffe on the other hand was suffering from a chronic shortage of fuel, spares and men. The weather turned foul grounding almost everything, including the Allies. This in effect was beneficial for *JG 400* allowing it to carry out essential repairs and revitalise the crews and ground staff. This was aided by the return of *Hauptmann* Wolfgang Späte from Russia to take over as *Kommodore* of *Jagdgeschwader 400.* This was a very popular move and approved by everyone, giving a great morale boost to the men. In addition to this, 90 Me 163Bs had been delivered to Brandis at the end of December.

The New Year saw the Russians advancing faster than anyone had anticipated and a number of airfields were closed down in the face of the advancing Red Army. A number of the crews arrived at Brandis still ready to do battle. The training of pilots continued even though there was a desperate shortage of fuel, which limited their flying training. So desperate were the German, that in one incident one of the Komet pilots, who had had only two 'sharp' starts in the Me 163B, was made operational and sent on an intercept mission. He had attacked a bomber formation and had been driven off by a P-51 Mustang which followed him down. As he ran out of fuel and glided into land, the P-51 Mustang attacked and hit the little rocket

One of the largest hangars at Brandis used for jet aircraft seen here after a raid by Allied bombers.

plane. The Komet made a perfect landing, but when the ground crew got to the aircraft they found the pilot, Feldwebel Herbert Klein, dead in the cockpit with a bullet hole in his head. The bullet had passed right through his seat armour.

By February the writing was on the wall and almost all Germany's military units were in a state of withdrawal. The *Oberkommando der Luftwaffe* – OKL (Luftwaffe High Command) ordered that production of the Me 163B cease. The total number of Messerschmitt Me 163 aircraft built was 364, of these 279 were production models and only 70 of those ever saw combat.

2

THE BACHEM BA 349 'NATTER'

This was one of the most unusual aircraft built during the Second World War. Born out of desperation, the Ba 349 'Natter' was designed to be used against the ever-increasing heavy bomber raids that were pounding the Third Reich. The theory was that the Natter would be launched from a vertical railed launching ramp and accelerated to a speed of around 400 mph. As the little rocket-powered fighter approached the bomber formation, the pilot would level the aircraft and attempt to manoeuvre behind one of the bombers. When the aircraft closed to around 200 yards, the pilot would then ripple-fire a battery of unguided rockets in one single attack. He would then manoeuvre the aircraft away using the last of his fuel until it was expended, before putting the aircraft into a gliding descent until he reached an altitude of 4,500 feet. Then, in theory, he would release his seat harness and the clips that held the nose of the aircraft in place. Hopefully the entire nose section, including the windshield, instrument panel and rudder pedals would then fall away, leaving the pilot in his seat open to the elements. As the nose section left the aircraft it activated a large recovery parachute that was fixed to the rear section of the fuselage slowing down what was left of the aircraft sufficiently for the pilot to fall out and descend to the ground by parachute. Parts of the aircraft that floated to the ground – the rocket motor and guidance system – were all re-usable.

The proposal for such an aircraft had come earlier in the war from Werner von Braun, but this had been rejected by the RLM as being unworkable because of the size of his design. Then *Dipl-Ingeneur* Erich Bachem of the Fieseler Company took a look at the idea and produced a cut-down version of the design, but that, like von Braun's, was rejected. Then in 1944 the RLM issued a requirement for a small interceptor that was cheap, simple and quick to produce. A number of companies, Heinkel, Junkers and Messerschmitt submitted designs, as did Fieseler. The aircraft selected was the Heinkel P.1077 rocket fighter; but Bachem had an ace up his sleeve. One of Bachem's friends was the SS *Reichsführer* Heinrich Himmler, who was always on the lookout to extend his power into the armed forces. Bachem persuaded Himmler to invest in an air defence weapon of 150 semi-expendable rocket fighters. This in turn caused the Luftwaffe to look again at the proposal because the last thing they wanted was the SS gaining a foothold in the air

Bachem Ba.349 vertical take-off fighter on the launching gantry.

Another view of the Bachem Ba.349 being prepared for an unmanned test flight.

Bachem Ba.349 being fuelled with C-stoff.

Bachem Ba.349 being recovered after its first unmanned test launch.

defence of Germany. The Luftwaffe ordered a further 50 from the Fieseler Company and by so doing gave official backing to the project.

A small factory was requisitioned at Waldsee in the Black Forest and detailed design work began in August 1944. The construction of the aircraft consisted of a wooden

Close-up shot of Oberleutanant Lothar Siebert climbing into the cockpit of the Bachem Ba.349.

A more distant view of Oberleutnant Siebert climbing into the cockpit, giving a better impression of the size of the aircraft.

Bachem Ba.349 'Natter' being launched.

The Bachem Ba.349 rolling onto its back before plunging to the ground killing the pilot Oberleutnant Lothar Siebert.

Oberleutnant Lothar Siebert had talked things over with Erich Bachem prior to that fatal test flight. What questions had he asked?

Flugkapitän Zacher (right) talking with Natter engineers about the Ba.349 M8 in the background.

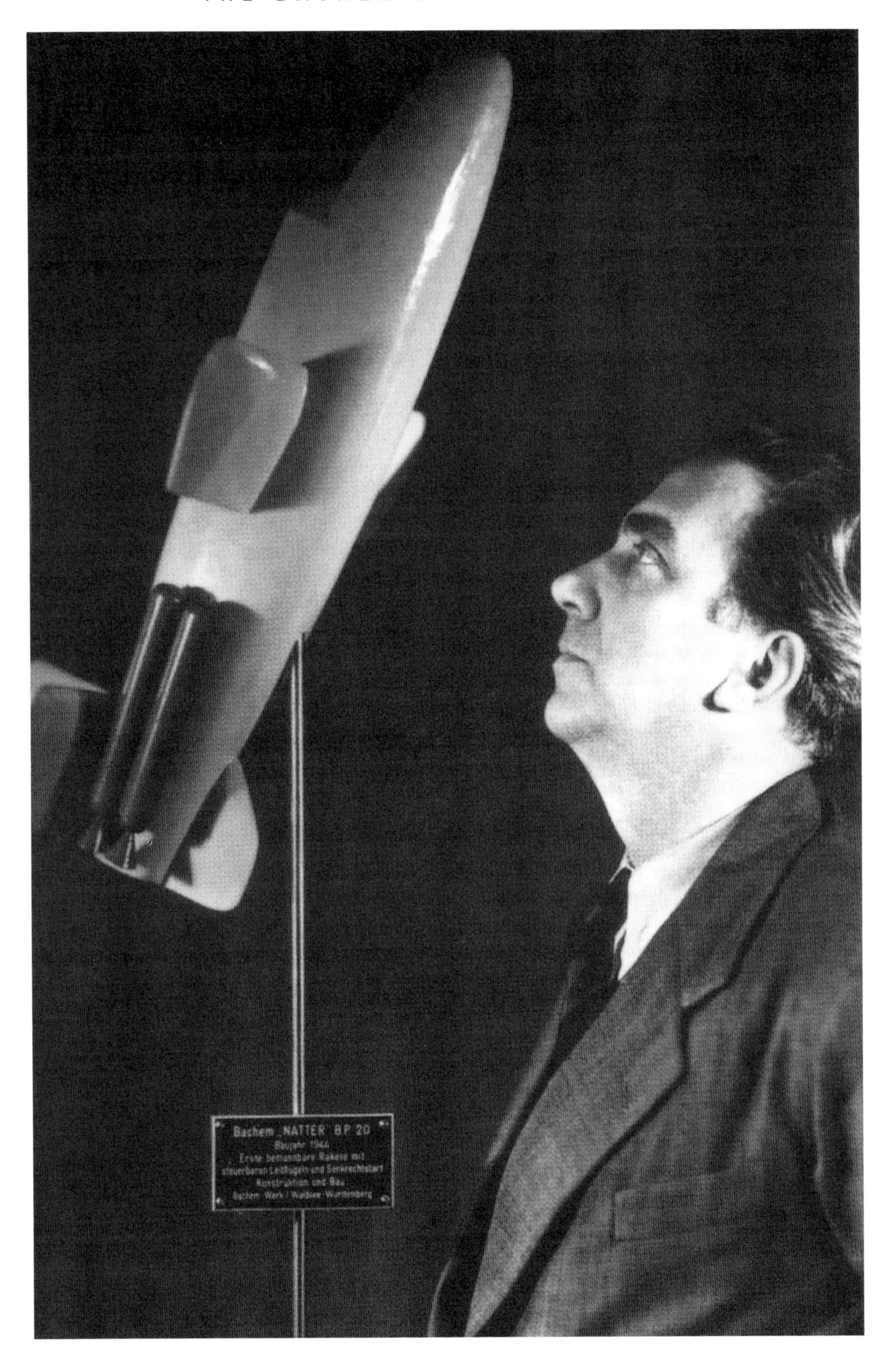

Dr. Erich Bachem with a model of his rocket aircraft, the Ba.349.

airframe, a cruciform tail section and a 19 ft. 9in. wing that had no movable parts and was constructed in one piece. The only metal pieces used in the construction were on the load bearing sections, hinges and the control push rods. The Natter was powered by a single Walter 109-509A-2 liquid fuel rocket motor that developed 4,400 pounds of thrust. The two fuel tanks were mounted in the fuselage, one containing 96 gallons of T-Stoff, the other containing 42 gallons of C-Stoff. In addition to the rocket motor, four Schmidding solid-fuel rocket boosters were fitted to the rear fuselage for additional thrust on take-off. These burned for just ten seconds after which they were released and the aircraft climbed away on the Walter rocket motor.

Vertical Launch

The Natter was launched in a vertical position on a 75 ft., three-channelled rail that held each wingtip and the lower fin of the tail section. On launch the aircraft was placed on autopilot for the first part of the climb and then reverted to the control of the pilot for the remainder of the flight. The sense of urgency instilled into the project ensured that the first of the aircraft was ready for flight-testing at the beginning of November 1944. The first tests were carried out as an unpowered glider towed by a Heinkel He 111 and flown by *Flugkapitän* Zueber. The aircraft was towed to a height of 18,000 feet and released. Throughout the descent the pilot found the little aircraft extremely stable and at 3,000 feet he initiated the escape procedure and both pilot and aircraft floated to the ground on parachutes.

It wasn't until 22 December that the first vertical launch was carried out, and then the aircraft was only fitted with the four solid-fuel booster rockets. The problems that had dogged the Messerschmitt Me 163B manifested themselves once again. The Walter 509 rocket motor was not ready and it was not until the beginning of February 1945 that the first one arrived. It was immediately installed and on 25 February the first launch took place, with a dummy pilot. The test was a complete success and both aircraft and dummy were successfully recovered. Three days later the first manned flight took place flown by Oberleutnant Lothar Siebert.

With the tiny little fighter perched precariously on the vertical launching rail, the rocket motor and the solid-fuel rockets were ignited and the Natter shot up the rail and into the sky. Suddenly observers saw the canopy fly away from the aircraft, and then the aircraft continued to climb but began arching onto its back. The aircraft continued to arc all the way to the ground with the pilot still strapped inside. The aircraft disappeared in a blinding flash followed by thick black smoke. The pilot did not survive. Despite this loss, manned tests continued and were successful. Then it was decided that it was time for the little interceptors to become operational and ten of the Bachem Ba 349s were assigned to an operational air defence site at Kircheim, east of Stuttgart.

For over a week the pilots and their diminutive aircraft stood by waiting to intercept the next heavy bomber formation that came their way. The only thing that came their

The Ba. 349 being fuelled.

The nose section of the Bachem Ba.349 Natter being removed for servicing.

Bachem Ba.349 found in hangar after the war.

way was an American tank unit, so the ten Natter rocket fighters were destroyed to prevent them falling into the hands of the Allies.

In total 36 Bachem Ba 349 Natter rocket interceptor fighters were built: of these 22 were gliders or used for testing purposes, ten had been blown up and three were captured, two by the Americans and one by the Russians. Whether or not the Natter would have proved to be viable will never be known, but once again it showed that ingenuity could be born of desperation.

3

FIESELER FI 103R REICHENBERG

This was perhaps an even more perfect example of desperation – a piloted flying bomb. In May 1944 the 5th *Staffel* of *II/Kampfgeschwader 200* also known as *Kommando Lange* after its *Staffelkapitän*, carried out a series of experiments with Focke-Wulf Fw 190s. The idea was that the Fw 190 would be fitted with the maximum size bomb the aircraft could carry, the pilot would aim the aircraft and the bomb at the target and then bail out. The chances of the pilot attaining any accuracy with this were tenuous. Then a conversion of the Fieseler Fi 103 (V1) was proposed, but was initially rejected in favour of using a glider-bomb version of the Messerschmitt Me 328. Fieseler had been a pilot in the First World War and had gone into aircraft production on leaving the German Army Air Force.

Dornier DO 17 with Lorin Ramjet on top carrying out tests.

Lorin Ramjet being fired whilst mount on the back of a truck.

A number of Me 328s were ordered but it soon became obvious that they were not going to be ready in time to make any substantial impact on the war. It was left to SS Colonel Otto Skorzeny to put pressure on the high command to go along with the suggestion of a piloted Fieseler Fi 103. At the time, Skorzeny, although only a Colonel, had more power than many of the Generals because of his close association with Adolf Hitler. Four design conversions were put forward by the DFS who had been given responsibility for the project.

Given the code name *Reichenberg*, the four designs were the Fi 103R-I, R-II, R-III and R-IV. Four flying bombs were taken to the Henschel factory and the first, the R-I, appeared just 14 days later. This was a single-seat model, fitted with skids and landing flaps with ballast in place of the warhead. It had no pulsejet fitted and was to be used purely in the test programme. The Fi 103R-I had a cigar-shaped fuselage constructed of pressed sheet steel with an aluminium-alloy nose cone. The tail section was also made of pressed sheet steel, but the wings were of a plywood construction that had a single steel-tube spar, which went through the fuselage centreline. The R-II was of similar construction but had an additional cockpit fitted in the nose for training purposes. The R-III was a single-seat model that was to be used as an advanced trainer and like the other models was fitted with a landing skid and landing flaps. It was powered by an Argus 109-014 pulsejet engine of 770lb thrust, which was mounted on top of the fuselage. The large intake was just behind the pilot's cockpit whilst the tail pipe extended over the top of the tail and beyond the rudder. The fourth model, the R-IV, had no landing aids and had a warhead mounted in the nose, but otherwise was similar in construction to the R-III.

The cockpit instrument panel of the Fieseler Fi 103 was of the simplest design, and consisted of a warhead arming switch, clock, airspeed indicator, altimeter, a combined turn indicator and inclinometer. Beneath the instrument panel and between the pilot's knees were a gyro-compass, a three-phase inverter and a battery. There was no radio installed, although during test flights with the Heinkel He 111 a communication system between the mother aircraft and the Fi 103 was fitted.

Dropping the Fieseler Fi 103 from a converted Heinkel He 111 over the airfield at Lärz was the initial test. The first two flights ended in crashes, so it was decided to enlist the help of two of Germany's top test pilots, Heinz Kensche and Hanna Reitsch both of whom were working for the DFS at the time. The two test pilots carried out a number of tests using R-I and R-II and although there were a few hair-raising moments, all the flights were completed. It has to be remembered that the initial tests on these 'aircraft' were carried out as glider tests.

Initial construction of the R-IV model was carried out in an assembly plant near Dannenberg and later at a secret factory known only as the Pulverhof VI assembly plant. A total of 175 of the R-IV models were built at the two factories. The R-IV differed from the standard flying-bomb (V-1) inasmuch as ailerons were added to the trailing edges, which were operated concentionally by the stick and rudder in the cockpit.

Self Sacrifice

The initial idea of a suicide-pilot or *Selbstopfermänner* ('self-sacrifice man') was dismissed from day one, although Adolf Hitler by this time was considering almost anything. The idea was that the pilot would aim the Fi 103 IV at the chosen target and as he closed upon it he would release the cockpit canopy, and then bale-out over the side using his back-parachute. Unsurprisingly, there were problems. When the canopy was released it had to hinge over 45 degrees before it would fall away, but the rear end of the canopy would hit the cowling on the pulsejet intake, thus preventing the canopy from opening fully. In addition, the speed at which the R-IV would be travelling would almost certainly prevent the canopy from opening. In short the chances of the pilot surviving were virtually nil and this was accepted by almost everyone. One suggestion put forward was to have an escape hatch beneath the pilot's seat but this was dismissed as being too complicated to operate efficiently.

The Japanese version, the Ohka, had the pilot sealed into the cockpit so there was never any thought given to the pilot getting out. There was obviously a subtle difference between the term *Selbstopfermänner* and *Kamikaze*, the latter, 'Divine Wind', (or more accurately *tokubetsu kogeki tai*, 'Special Attack Unit' – the word *Kamikaze* was adopted by the Allies) most certainly meant suicide pilot.

These problems aside, there were over a thousand who volunteered for such missions and 70 were finally selected for training, though not one ever actually flew a mission. For a variety of reasons the *Reichenberg* Plan was never treated seriously enough to be carried through and when General Werner Baumbach took over as *Geschwader-Kommodore* of *KG 200* in October 1944 the whole programme was allowed to fade into memory.

The unmanned V-1, as it became known, inflicted a large number of casualties in Britain and its drone as it passed overhead, followed by the silence as the engine cut out, instilled fear into the hearts of all those who heard it.

Fieseler Fi 103 (Vergeltungswaffe retaliation weapon. aka 'doodlebug') being readied for launch.

Fieseler Fi 103R, the intended piloted flying bomb that became a terror weapon.

Fieseler Fi 103 being placed on a launching ramp.

Fieseler Fi 103 'doodlebug' about to crash in a residential district of London.

Fieseler Fi 103R Reichenberg after being captured by British troops.

Werner Baumbach shelved the Fieseler Fi 103 Reichenberg programme in favour of the *Mistel* (Mistletoe) project in October 1944.

Piloted version of the Fieseler Fi 103R.

British soldier trying the Fieseler Fi 103 Reichenberg on for size.

4

MESSERSCHMITT ME 328

Although appearing to be a follow-on to the Fieseler Fi 103, the Me 328 was actually devised at the end of 1942. The idea had first inspired the production of a cheap single-seat fighter that could be utilised either as a low-level bomber, or in an emergency, a fighter. It had been originally designed as a 'parasite' fighter to protect bomber formations but during the design phase a number of other uses for the aircraft had been put forward.

The aircraft was designed as Messerschmitt project P.1073 in 1941 and handed over to the *Deutsche Forschungsanstalt fur Segelflug* (DFS – German Research Institute for Sailplane Flight) for development. The job of building the aircraft was given to the Jacob Schweyer *Segelflugzeugbau* Company who used a variety of composites for the test frames, although the finished aircraft would be built largely of wood. Gliding tests were carried out using a Dornier Do 217E, together with launch tests from a catapult on the ground. These were followed by powered tests, the aircraft being fitted with two Argus As 014 pulsejets, the same as were used on the V1 flying bomb. The violent vibration created by the pulsejets proved to be too much for the Me 328's light wooden airframe, and two prototypes were destroyed.

The Me 328 was of a standard mid-wing configuration with a circular sectioned fuselage. The cockpit was raised with the rear of the canopy moulding into the fairing that tapered

The AS 014 pulse-jet mounted beneath the wing of Me 328A V2.

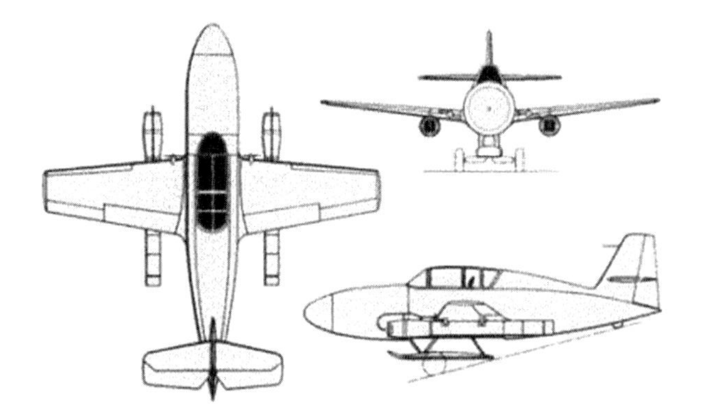

Three-view drawing of the Me 328 with pulsejets mounted beneath the wings.

Dornier Do 17 with an Me 328 mounted on top.

The all wooden Me 328 VI.

back to the tail section. The single fin of the tail had the tailplane fitted halfway up. The wings had inboard landing flaps and leading edge slots, although the test models had detachable tips that enabled the wingspan to be altered. The fuel tanks, which were self-sealing, were housed in the rear section of the fuselage and in the nose. Pilot protection was a bulletproof windscreen and armour plating around the seat. The undercarriage consisted of a retractable skid to which a dolly was fixed for take-off.

Although the Jacob Schweyer *Segelflugzeugbau* was making the aircraft, tests with models were being carried out at the Messerschmitt wind-tunnel at Augsburg to find the best position to place the two Argus pulsejet engines. The initial design placed the engines either side of the fuselage behind the cockpit with the tailpipes extending beyond the tail and mounted on small outriggers. After tests it was found that the physical vibration caused by the engines seriously weakened the outriggers rendering them unstable, although it would have reduced the noise levels considerably for the pilot.

After a number of tests where the wing positions were altered, it was decided that the best position to mount the engines was to place one under each wing, with the

Early version of Me 328 prototype mounted on a Dornier Do 17.

Messerschmitt Me 328 with pulsejets mounted on either side of the fuselage.

tailpipes terminating just under the tail section. The outriggers for this modification were considerably stronger and far easier to install, although later it was discovered that the pulsating exhaust of the pulsejet engine affected the rear portion of the fuselage and tail section.

The early work on the pulsejet engines had been carried out by Paul Schmidt of the Argus Company and, after negotiations, the Me 328 design team, under Dr. Ing. Fritz Grosslau, were allowed to use one of his early models in their tests.

There were two versions built, the Me 328A by Jacob Schweyer *Segelflugzeugbau* and the Me 328B by the *Deutsches Forschungsinstitut für Segeflug* (DFS). The basic idea of the programme was to build a fast, disposable fighter that would cost less than a quarter of what it took to build either a Messerschmitt Bf 109 or a Focke-Wulf Fw 190.

Like the tests of the Me 163 Komet, the first flights were made as gliders after being dropped from a Dornier Do 217E (JT+FL) at Hörsching, Austria. The drops were made from varying heights, but the overall handling results were found to be very poor. For the purposes of using the aircraft as a piloted-bomb the results showed that it was just about aerodynamically acceptable if the pilot baled out late enough.

The first of the powered flights using either the three-wheeled *Lippisch-Latscherwagen* with the Madelung KL-12 cable-type catapult or by an Rh B rocket-propelled rail carriage did nothing to improve the aircraft's overall performance and a number of accidents occurred as a result of structural failures in the rear section of the fuselage and tail because of the pulsejet acoustics. Air launches were carried out by either towing the aircraft, or by using a launch rig fitted on top of the launch aircraft. If the aircraft were to be used as fighters against bombers, the fixed towing method was the one that had been selected as the fuel saved gave the aircraft more time on station.

The first production model was to be the Me 328A-1 which was to have the two pulse-jets mounted beneath the wings and which was detailed as a hook-on fighter with the Heinkel He 177 bomber to act as escort when needed. The second model, the Me 328A-2, was intended to have four fuselage-mounted pulse-jets and, like the A-1, was to be towed by the four-engined Messerschmitt Me 264 heavy bomber as an escort fighter. After a series of tests it was discovered that the maximum operating ceiling was well below that of the Allied fighters. It was obvious that this was a no-go project and the A-series was dropped.

The information gained during the testing of the Me 328A was absorbed into the Me 328B programme, a fast-bomber project using the pulse-jet. This was to be a glider-escort project where the aircraft was mounted on top of the fuselage of either Junkers Ju 88s or Ju 388s. After numerous tests it was decided that the whole project was impracticable and it was abandoned. There was one further model, the Me 328C powered by a Junkers Jumo 0048 turbo-jet. The intention was to install the engine inside the fuselage of the aircraft, but the cost of carrying out this project went against the original principle of a cheap aircraft. Late in the war, the design was resurrected for consideration as a *sellbstopfer* (self-sacrifice) aircraft, but was judged unsuitable even for this purpose.

Nr.5/KG 200 ordered the unpowered version of the Me 328B as a piloted glider-bomb, but none were ever forthcoming. The project was abandoned despite production facilities being readied to construct the aircraft at the Jacob Schweyer sailplane factory.

5

HEINKEL HE 178 & 280

Although there had been a number of attempts to 'marry' a turbojet engine to an aircraft, the first to claim success in that field was Professor Ernst Heinkel, and the first jet aircraft to take to the air was the Heinkel He 178. The aircraft had started life as a test bed for Professor Ernst Heinkel's jet engine, the HeS 3b, which was a development of the HeS 3a engine. This was a private venture for Ernst Heinkel that had started back in March 1936, when 24-year-old Hans-Joachim Pabst von Ohain, a protégé of Professor Pohl of the University of Göttingen, joined the Heinkel company after a request from Professor Pohl for von Ohain to be given the chance to develop his concept of jet propulsion. Heinkel agreed, remembering fondly the time when he was the same age as von Ohain and had been given the chance to spread his wings, so to speak, by the LVG (*Luft-Verkehrs Gesellschaft)* Company.

The reason that it had to be a private venture was that the RLM (German Air Ministry) would not give Ernst Heinkel a single Reichsmark towards any development projects that

Heinkel He 280 taking off on a test flight.

Heinkel 178 V2 outside the factory.

were based on theory alone. It also had to be kept secret because any number of other aircraft manufacturers would only be too eager to 'steal' the information. Heinkel set up a workshop on the other side of the airfield at Rostock-Marienehe, well away from prying eyes, and left von Ohain to conduct his experiments. Within a matter of months he had registered the patent for an axial turbojet fitted with a centrifugal compressor and an inward radial turbine.

Another test, this time by Werner von Braun, took place in March 1937 using a Heinkel He 112 V5. Ernst Heinkel offered the services of one of his best engineers, Walter Künzel, along with a team of his best riggers, to von Braun back in 1935 to help him with his rocket research. In addition he had offered von Braun the use of the Heinkel He 112. This was accepted and on the first trial the rocket motor was fired, but blew up. Fortunately it was fired by remote control so no one was injured. A second attempt with another He 112 suffered the same fate, but after major modifications it was ready for the first flight. After first warming up the Jumo 210 piston engine, the pilot, Erich Warsitz, started the rocket engine. As he eased the throttle forward to taxi out onto the runway, there was an almighty bang and the rocket engine exploded. The Jumo 210 engine flew into the watching spectators, whilst the pilot was blown from the aircraft. Miraculously, there were no serious injuries to the spectators or to the pilot.

Heinkel supplied a second aircraft, only this time the aircraft took off under the power of the standard engine, which once airborne, was shut down, and the rocket engine fired. The tests were a complete success for von Braun. In the meantime Ernst Heinkel returned to his own experiments.

In September 1937 Pabst von Ohain fired up his hydrogen-fuelled engine for the first time. The HeS 1, as it was designated, more than lived up to expectations; but despite claims by the Germans that this was the first jet engine, it wasn't. In England Frank Whittle had fired up his jet engine five months earlier. Whittle was not able to progress much further because of the usual British reticence to push forward with new ideas and it was to be June 1944 before the first RAF squadrons were equipped with the Meteor jet fighter. Despite the fact that his first engine was inferior to that of Whittle's, von Ohain persevered and soon produced the next version, the HeS 2. This engine produced an initial thrust of just 200lb, but after some major modifications produced a thrust of 800lb and was redesignated the HeS 3. Impressed by the bench tests, Ernst Heinkel authorised air tests to be carried out using the design of this engine. The Heinkel He 118 was selected as the aircraft to be fitted with the HeS 3a engine.

The He 118 was one of four designs that Ernst Heinkel had submitted to the RLM in their request for a high-performance dive-bomber. It had first flown in June 1937 and was an all-metal, low-wing monoplane powered initially by a Rolls-Royce Kestrel engine. The second improved prototype was the one chosen to carry the HeS 3a engine. The aircraft was powered by a Daimler Benz DB600 engine and carried the HeS 3a engine slung beneath the fuselage. The first tests took place at the beginning of May 1939, just one month before the He 176 took to the air.

Ernst Heinkel was, at the time, amongst those in the forefront of rocket research and was developing the Heinkel He 176, the first rocket-powered aircraft. He had successfully experimented in 1936 with a Walter rocket motor attached to a Heinkel He 72 biplane.

The initial tests of the He 118 were carried out and the reciprocating Daimler Benz DB600 engine was cut out for lengthy periods of time whilst the HeS 3a turbojet engine flew the aircraft. Satisfied with the results, Ernst Heinkel set to work designing an aircraft around the HeS 3b engine, an uprated version of the HeS 3a. So secret was the work that the RLM knew nothing of its existence until the end of 1938, when they were told by Ernst Heinkel of the research work that had been carried out on a revolutionary form of jet propulsion. As typical bureaucrats they showed almost no interest in the project, possibly because their noses were out of joint as they had been kept in the dark until the very last moment and not been consulted at any time. They were also concerned that Heinkel, who in their opinion was a builder of aircraft, was now getting involved in the development of engines.

Heinkel 280 taking off on a test flight with its engine covers removed.

Heinkel He 178, the first turbojet aircraft.

They would have preferred engine manufacturers to become involved in the progressive technologies. Assistance was further hampered by the fact that *Generaloberst* Ernst Udet and *Generalfeldmarschall* Erhard Milch were both of the opinion that the jet fighter was superfluous to the Luftwaffe‘s needs. This, coming from two pilots whose flying and combat experience was limited to flying biplanes of a bygone age, was not exactly encouraging. Unfortunately, they both evidently held very high ranks in the Luftwaffe and both had the ear of *Reichsmarschall* Hermann Göring.

Heinkel ignored them and instructed Robert Lüsser, who was head of the design bureau, to concentrate on the design and development of a jet fighter, the Heinkel He 180. Aircraft manufacturers Messerschmitt and Henschel were also developing their own jet aircraft, but it was Ernst Heinkel who was ahead of the game. The aircraft developed for the HeS 3b was the He 178 V1 (DL+AS – this was the number on the fuselage). It had a simple monocoque-designe fuselage made of duraluminum, with the pilot's cockpit placed well forward of the leading edge of the wings, which were of a predominantly wooden construction. The engine was mounted inside the fuselage with the front-end level with the trailing edge of the wings. The engine's air intake was at the nose of the aircraft and was channelled through a duct that curved its way through the fuselage, under the pilot's seat and fuel tank, into the engine. The tailpipe, which was straight and long and almost one-third of the aircraft's length, exited beneath the tail.

RLM Reluctance

The first tests took place at the beginning of August 1939 and consisted of a series of taxiing trials. During one of the tests, on 24 August, the pilot *Flugkapitän* Erich Warsitz made a short flight along the runway. It wasn't until 27 August that the aircraft took to the air properly and not without incident. The take-off was textbook but the undercarriage refused to retract. The test flight continued around the airfield's perimeter, then on the second pass a bird was sucked into the air intake causing the engine to cut out. Fortunately Erich Warsitz was able to make a safe power-off landing and very little damage was caused. This was the first flight of a jet-powered aircraft and it was to be another year before the next one – surprisingly, that would be made by the Italian Caproni Campini CC.2.

Ernst Heinkel, now convinced that his aircraft had potential, approached the RLM for support. Once again their reaction was less than enthusiastic which, considering the political climate in the autumn of 1939, was understandable. Despite this Heinkel kept up the pressure and was rewarded by the RLM agreeing to a demonstration. On 28 October *Flugkapitän* Erich Warsitz performed the first public flight of a jet aircraft in front of a number of very senior Luftwaffe officers, who included Ernst Udet and Erhard Milch. Having persuaded the RLM to sanction the demonstration the resulting display was hardly earth shattering. The projected performance of the aircraft was far greater than the reality. The theoretical maximum speed of the He 178 V1 (DL+AS) was estimated to have been 435 mph, but in the demonstration the maximum speed achieved was only 185 mph. This was put down quite simply to the limited power of the engine and the basic design of the aircraft. There was a second model, the He 280 V2 (GJ+CA) and although powered by the same engine it had a larger wingspan and wing area.

The lukewarm reception that Ernst Heinkel received did nothing to dampen his enthusiasm, in fact quite the reverse. It was blindingly obvious that the fighter required a much more powerful power plant and Heinkel set about the task of developing one. He was helped when the Junkers engine division, which had merged with the Junkers Aircraft Company in 1936, took over the secret development projects at Magdeburg. This left an engineer by the name of Max Adolf Müller, who headed up the project department at Magdeburg, free to join the Heinkel Company. Müller had been working on an axial turbojet engine, the Jumo 109-006, and on joining Heinkel brought all the information with him. In the meantime Dr. Pabst von Ohain continued to develop the HeS 3b engine and produced an uprated version, the HeS 6. It was installed into the He 178 V2 (GJ+CA) but did little to improve on the mediocre performance of the He 178 V1 (DL+AS).

Rear three-quarter view of Heinkel He 280.

Front three-quarter view of the Heinkel He 280 prototype.

The reluctant support from the RLM suddenly became more forthcoming when the head of the reaction engine department at the *Technische Amt*, Hans A. Mauch, was replaced by a younger and much more forward-looking engineer, Helmut Schelp. He pushed Ernst Heinkel into continuing with his development programme on the Heinkel He 280, motivated by the threat of war and the strong possibility of conflict with Britain.

The Heinkel 280 V1 (DL+AS) was an all-metal, mid-wing, monoplane that was powered by two turbojets mounted beneath each wing. The wings had straight leading edges but curved trailing edges and were attached mid-fuselage. The fuel was carried inside the fuselage, which also contained the armament in the nose. The cockpit, which was mounted just in line with the leading edge of the wings, had a sliding canopy. The tail unit was comprised of two fins and rudders mounted either side of a single elevator. The tricycle undercarriage consisted of two main wheels attached to the wings just inside of the engines (retracted partly into the wing and partly into the fuselage), whilst the nose-wheel retracted backwards into the nose section of the fuselage. The cockpit was also fitted with the world's first ejection seat that operated by compressed air and was designed with the idea of having cabin pressurisation introduced at a later date.

First Flight of the He 280

The first flight of the Heinkel He 280 V1 (DL+AS) was on 2 September 1940 as a glider, when it was towed by a Heinkel He 111B in a series of tests. Between September 1940 and the end of April 1941, 40 glide tests were carried out. The reason for the lengthy glide tests was due in part to the fact that the engines were not ready until the end of March 1941. Satisfied with the results, Ernst Heinkel decided that the HeS 8 (designated 109-001) engines, producing around 1,100lb of thrust, should be fitted and work began immediately. All turbojet engines were given the prefix 109, irrespective of who built them.

On 2 April 1941 the Heinkel He 280 V1 took to the air with test pilot Fritz Schäfer at the controls. This was the first time a purpose-built jet fighter had taken to the air and a

Heinkel He 280 prototype.

milestone in aviation history. This first flight was made ostensibly in private and carried out at just under 1,000 feet. The aircraft had no cowlings over the engines, this was because they had a tendency to leak fuel. The undercarriage was not retracted and the fighter made only one circuit of the airfield before landing.

Three days later the first official flight took place before a number of distinguished officials from the RLM and Luftwaffe that included Ernst Udet, Wolfram Eissenlohr, Ing. Reitenbach and Helmut Schelp. The demonstration went off without a hitch and left the RLM officials in no doubt that work on the jet engine should be intensified.

Work began almost immediately on a second version of the HeS 8 (109-011), which was proposed to have a static thrust of 2,866lb. At the RLM's 'suggestion' the specialist engine manufacturer of Hirth Motoren GmbH was placed under the control of the Heinkel factory. The Hirth Motoren factories in Berlin and Stuttgart-Zuffenhausen specialised in the production of reciprocating engines and turbo-superchargers. The idea was to bring the entire specialised skilled labour under one roof, so to speak.

In the meantime, Max Adolf Müller had continued with his work on the HeS 30 (109-006) engine and had been moved to the factory at Stuttgart-Zuffenhausen. This left Professor Hans-Joachim Pabst von Ohain free to continue his development of the HeS 8 (109-011). Problems began to develop with the HeS 8 with regard to the engine's size. Von Ohain had been striving to keep the diameter of the engine as small as possible, but with increased thrust. It wasn't working and the design of the Junkers Jumo 109-004 had overtaken it. The slightly improved HeS 8 (109-011) engines with 1,323lb of thrust were fitted into He 280 V2 (GJ+CA) and HeS 280 V3 (GJ+CB). The results were not encouraging and it was also discovered that the He 280 had developed a tail flutter. There were problems with the engine as well, and it was the failure of a turbine blade that caused the V3 (GJ+CB) to have a serious crash landing.

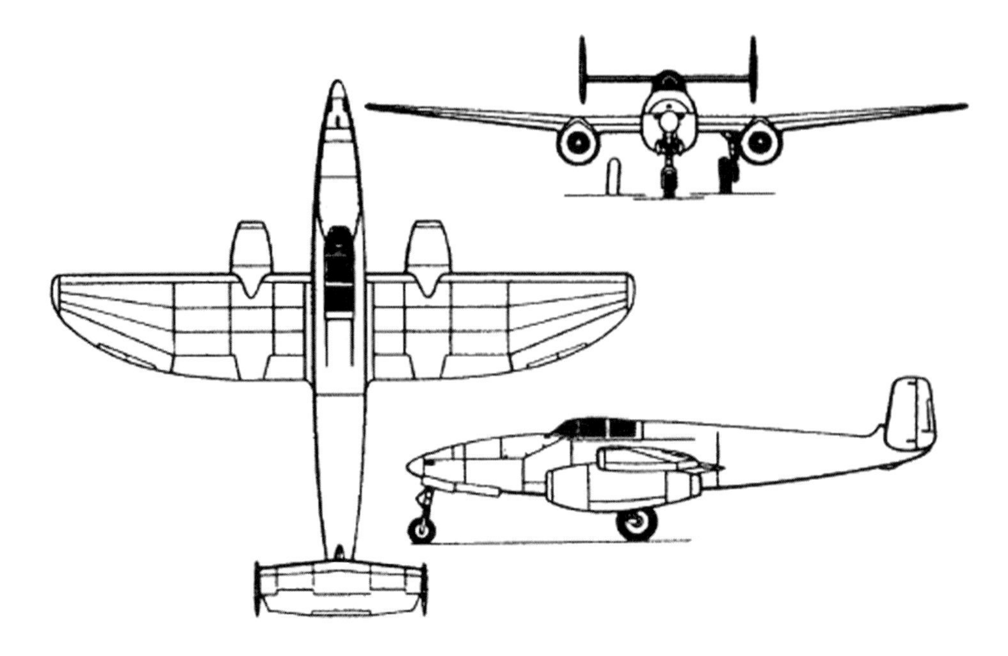

Three-view drawing of the Heinkel 280.

In Stuttgart, Max Müller's development of the HeS 30 (109-006) was showing tremendous promise, so it was decided to abandon the development of the HeS 8 (109-011) and concentrate on the HeS 30 (109-006). Work would continue to a smaller degree on the HeS 8 (109-011) alongside the work on the HeS 30. For some inexplicable reason Helmut Schelp decided, for reasons known only to himself, that it was the development work on the HeS 30 that should be discontinued and that all efforts should be channelled into developing the HeS 8. This in effect left the Heinkel He 280 without an engine. By the time the decision to discontinue with the HeS 30 had been taken, Max Müller had left the company, possibly because he could see all his efforts being eroded.

The Heinkel He 280 V2 (GJ+CA), the (V3 (GJ+CB) having been destroyed in the crash landing), was fitted with the Junkers Jumo 109-004 engines but on the first flight the aircraft was destroyed after the undercarriage collapsed on landing. Development of the aircraft continued with the He 280 V4 (GJ+CC), which was built using the BMW(*Bayerische Motoren Werke*) 109-003 engines. On 15 August 1943 the aircraft took to the air and carried out all tests successfully. But still there were nagging doubts about the power-to-thrust ratio, which were proved right when the V5 (GJ+CD) was fitted with the HeS 8 engines. These were quickly removed and the aircraft re-engined with the BMW 109-003 engines. The He 280 V6 (NU+EA) was also fitted with the BMW 109-003 engines and together with the V5 (GJ+CD) was being used in armament trials carrying out mock combat exercises with the Focke-Wulf 190. These trials showed that the jet-powered fighter was superior to the piston-engined fighter in both speed and manoeuvrability. It was also thought by many to be a better fighter than its rival at the time, the Messerschmitt Me 262, but it has to be remembered that these trials were carried out for very short periods of time and not under battle conditions. In the end the

Heinkel 280 was phased out in favour of the Me 262. Some say it was a political decision, but in reality the Me 262 had a longer range, was far better equipped with cannon and machine guns and was a sturdier machine. The official reason given was that the company had only been granted permission for nine prototypes to be built, and in their opinion there were too many problems with the aircraft for them to be resolved in such a short time.

The Heinkel 280 V7 (NU+EB) was transferred to the DFS (*Deutsche Forschungsinstitut für Segelflug* – German Institute for Glider Flight) and the engines removed. It was used to examine a variety of wings and tail sections and in all carried out 115 flights towed behind a Heinkel He 111H-6. An eighth Heinkel 280, the V8 (NU+EC), which was fitted with Junkers Jumo 109-004 engines, carried out a number of tests successfully. On 29 September it took to the air with a new V-type tail section, replacing the twin fin and rudders of the previous models. The information gained in these trials was instrumental in helping to design the Heinkel He 162 tail assembly.

There was one more Heinkel 280, the V9 (NU+ED). This was used as an engine test bed for the BMW engines and flew on 31 August 1943 with BMW 109-003 engines installed.

6

HEINKEL HE 162

With the relative success of the Messerschmitt Me 262 and the Me 163B and the demise of the Heinkel 280, Ernst Heinkel took advantage of a requirement by the RLM for a Volksjäger (People's Fighter). Towards the end of 1944 and with events in Europe not going the way the Third Reich had intended, the RLM underwent a major re-organisation. All departments concerned with the production of aircraft and anything associated with it were transferred to Albert Speer's Ministry for Armament and Ammunition. Within this organisation the technical department, under the control of Karl Otto Saur, answered directly to Albert Speer. The desperate need for raw materials was of the highest priority to Albert Speer and obtaining supplies was becoming increasingly difficult. Saur came up with a concept for a new lightweight jet fighter which would use the minimum of essential raw materials, and yet still be superior to any of the piston-engined fighters in service at the time. The fighter could not weigh any more than 4,410lb, would have a short take-off capability of a minimum of 1,640 ft and would carry armament of twin 30-mm cannons. The manufacturer selected would have to have the aircraft ready by 1 January 1945, which was less than four months hence.

Out to Tender

Albert Speer agreed to the idea and on 8 September 1944, Saur issued a requirement to the aircraft manufacturers Arado, Blohm & Voss, Focke-Wulf, Junkers, Heinkel and Messerschmitt. Messerschmitt almost immediately expressed no interest in the project as they had more than enough to do in producing their existing jet fighter programme. Focke-Wulf on the other hand put forward a totally unrealistic proposal, whilst Arado's project proposal was rejected out of hand. Even Heinkel's proposal was rejected initially, but one of Heinkel's directors Dr.Ing. Franke complained bitterly that all the other manufacturers had calculated the weights and performance of their aircraft using a completely different formula to that of Heinkel's. The following day the other manufacturers calculated their weights and performance figures using the same formula as Heinkel's. New proposals from the various manufacturers resulted in a number of designs being put forward and the two accepted for further consideration were the P.211 of Blohm & Voss and the P.1073 design of Heinkel. After a heated debate no decision was reached, but on 23 September 1944 the Heinkel team at the Vienna-Schwechat factory put together a mock-up of the proposed aircraft. It was inspected by General Lucht and a number of other officials from Speer's department and approved, despite the general consensus that the Blohm & Voss design was infinitely superior. Such was the desperation for a new fighter that Göring approved the contract the very same day and the Heinkel team set to work.

Because of the gathering storm regarding the selection of the aircraft it was decided to hold one more meeting at the RLM, where once again heated arguments broke out amongst the parties whose aircraft had not been selected. To solve all the arguments as to whether or not the choice should be the Heinkel or Blohm & Voss, the two representatives from *Aerodynamischen Versuchsanstalt*, Professors Betz and Küchemann, were asked to make an informed decision. Once again politics prevailed and a half-hearted statement that the intake ducts on the Blohm & Voss design might cause problems was enough to close the arguments as far as the RLM were concerned.

Whilst these tests were being carried out to try and improve the aircraft, in Berlin Hitler, now realising that the German Army had its back against the wall but still believing that he could not be defeated, issued a decree calling for the formation of the *Volkssturm* (Home Guard). It was to be under the direct control of *Reichsführer*-SS Heinrich Himmler.

Hermann Göring, slowly losing favour, set up meetings with *Generaloberst* Keller, Chief of the National Socialist Air Corps and with *Reichsjugendführer* Artur Axmann, leader of the Hitler Youth Organisation, to consider setting up *Volksjäger* units to operate alongside the *Volkssturm.* Axmann, eager to please Göring, assured him that the Hitler Youth Organisation

Heinkel He 162 taking off.

had the manpower required to fulfil any requirement for pilots that Göring needed. The two of them agreed that an entire year's intake would be sent first for gliding lessons and then sent on conversion courses to the *Volksjäger.* These young men were to fly jet fighters after training that could only have been measured in weeks and against battle-hardened Allied fighter pilots. A large number of the Luftwaffe's top pilots, amongst them the legendary *General der Jagdflieger, Generalmajor* Adolf Galland, were appalled and voiced their disbelief at the extravagant claims of Axmann and their disapproval of using inexperienced and untutored pilots. In any case, the *Volkssturmstaffeln*, as it would be called, would have to wait for the aircraft to be assessed by the Luftwaffe before it could be handed over.

The rift between Hitler and Göring became even more apparent when Hitler took the entire jet programme out of the hands of the Luftwaffe and placed it in the hands of the SS. The man put in charge was *Obergruppenführer* Dr-Ing Kammler with the title of 'General Plenipotentiary of the Führer for jet-propelled Aircraft'. Göring, not to be seen taking a back seat, appointed his own 'Special Plenipotentiary for Jet and Rocket Aircraft', the former *General der Nachtjagd, Generalmajor* Josef Kammhuber. All this did was to confuse everybody as to who was in charge. All the time the Allied forces were closing in and the training programme at Marienehe had to be abandoned because of the rapid advance of the Russian Army.

Even the maintenance of the *Volksjäger* was now being regarded as unnecessary, as it was envisaged that thousands of He 162s would be coming off the production line and it would be a waste of time repairing or maintaining them.

After working day and night, the Heinkel design team, headed by Chief Project Designer Siegfried Güther and Chief Designer Karl Schwärzler, produced detailed drawings of the He 162A. The designation He 162 had been given to the aircraft by the RLM although Ernst Heinkel wanted the designation He 500. The construction of the *Spatz* (Sparrow) as it was called (the name was later changed unofficially to Salamander), was simplicity itself. The fuselage was of a semi-monocoque construction made of duraluminum with a moulded plywood nosecap. The nose-wheel door and radio compartments were made of plywood and the cockpit was equipped with the minimum of instruments. The shoulder-mounted wings had a wooden framework with an aluminium skin that was bonded to it with no access panels. The auxiliary spar in the wing carried the ailerons and flaps, which were hydraulically operated and could be depressed 45 degrees. Only the tips of the wings were metal, this allowed internal servicing using special tools equipped with mirrors. Four vertical bolts attached the wings to the fuselage whilst the turbojet, which was mounted on top of the fuselage behind the pilot, was fixed with three connections. Although the designers expressed the belief that the visibility was good, one has to accept that because of the engine being mounted on top of the fuselage, rear visibility was virtually nil. The narrow-track undercarriage retracted entirely into the fuselage hydraulically but was lowered by means of springs.

The tail section consisted of twin rudders that had a pronounced dihedral and a hinge at its forward end, for longitudinal trimming could alter the incidence of the entire tailplane. The twin rudders could be offset for directional trimming whilst in flight, but to adjust the trim tabs on the ailerons the aircraft had to be on the ground.

Prototype Heinkel He 162.

Heinkel He 162, the Volksjäger ('people's fighter').

He 162A taking off.

The Heinkel He 162 was orginally intended to be flown by the Hitler Youth because of the lack of experienced pilots, a fairly absurd idea for such a machine.

Brand new He 162A on a tail stand.

Heinkel He 162 of 'Wild Sau'.

Heinkel He 162 coming in to land after its first test flight.

The main fuel tank was situated behind the cockpit and contained 168 gallons (695 litres), whilst a further 39.6 gallons (180 litres) was carried in between the main and auxiliary wing spars. The He 162A also had one of the first ejector seats that was fired by means of a single cartridge. This too had been designed by Heinkel. Entry and egress of the cockpit were achieved by means of a fold-back canopy and the seat itself was fitted with a Sutton harness. In the well of the seat was an emergency oxygen system, the seat itself was made up by the parachute. The main oxygen supply came from a 3 -pint metal container mounted on the port side of the fuselage with an indicator on the instrument panel facing the pilot.

Flight instruments in the cockpit were kept to a minimum and consisted of an FK 38 magnetic compass, a fine-course altimeter, an ASI (Air Speed Indicator), a turn-and-bank indicator and a Revi 16G gunsight, which was mounted directly in front of the pilot. Armament consisted of two MG 151 20 mm cannons with 120 rounds per gun of ammunition. The bulge in front of the control column housed the nose-wheel when it was retracted. What is truly remarkable is that the aircraft was designed, built and flown all within the space of 90 days. By any standards that was an incredible achievement.

Into Production

Three main assembly plants were set up, two of which, Heinkel-Nord at Marienehe and Junkers at Berburg, were expected to attain a production rate of 1,000 aircraft a month.

Heinkel 162 during a demonstration flight with the starboard aileron starting to come off.

Heinkel 162 with the starboard aileron almost detached, causing the aircraft to go into a spin, killing the pilot.

Heinkel 162 on low level flight during a demonstration to senior Luftwaffe officers.

Heinkel He 162 touching down after a test flight.

The third plant at Nordhausen was scheduled for 2,000 aircraft per month. A network of sub-contractors would feed the plants and furniture making and woodworking companies in the vicinity of Erfurt and Stuttgart would manufacture all the wooden components. Heinkel at Barth, Pomerania; Pütnitz, Mecklenburg; Stassfurt, Saxony and Oranienburg, Berlin would produce the all-metal fuselages. The Junkers plants at Aschersleben, Leopoldshall, Halberstadt and Bernburg were tasked to make the fuselages. The BMW 003 engines were to be manufactured in an old saltmine near Urseburg. Other saltmines and even a chalkmine were brought into use as underground factories and the extremely tight schedule called for 1,000 He 162s to be ready by the end of April 1945.

The first of the prototypes, the Heinkel He 162 V1, with *Flugkapitän* Gotthold Peter at the controls, took to the air on 6 December 1944 from Schwechat. The flight lasted 20 minutes and attained a speed of 522 mph at a height of 19,685 feet. The flight had to be terminated when one of the plywood undercarriage doors came off. A second flight four days later ended in tragedy when, in front of a crowd of RLM, Luftwaffe and Nazi officials, the leading edge of the starboard wing ripped off during a high-speed low-level flight. The aircraft started rolling and seconds later the starboard aileron and wingtip flew off with the result that the aircraft crashed and the pilot was killed.

Urgent enquiries into the accident revealed that the bonding agent was defective and in order to reduce any disquiet about the accident Technical Director Dr.Ing. Francke announced that he would fly the next test flight in the He 162 V2. On 22 December 1944 the first flight of the V2 took place and Francke took it to its structural limits to demonstrate the confidence he had in the aircraft. Francke flew all subsequent test flights and it was flown to its limit every time. The V2 was then fitted with two MK 108 30-mm cannons and used for firing trials. Whilst these trials were going on, the information gathered from the flight tests of the V1 and V2 were used to develop the He 162 V3 and V4. Among the changes to both the aircraft was the bringing of the c.g. (centre of gravity) forward, and to do this a weight was fitted over the nosewheel. The tail sections were enlarged slightly and in order to reduce the effective dihedral angle, a pronounced anhedral was applied to the wingtips.

These four prototypes were regarded as pre-production models and as such a most complex numbering system was applied to them. Because the He 162 was also being built by other manufacturers and at different plants, the numbering system differed. The fighters being built at Schwechat, of which a series of ten had been ordered, were numbered He 162A-0, whilst at the nearby Hinterbrühl facility He 162A-1s were being produced. In the meantime the He 162 V5 had been built and assigned to static tests. This was quickly followed by the V6, which was flown for the first time on 23 January 1945, and which was the last to be fitted with the MK 108 cannon. The He 162 V7 was the first to be fitted with the He 162A-1 airframe built at Hinterbrühl and was used exclusively for vibration tests, which revealed the need for the structure to be strengthened. The following three, V8, V9 and V10, were all fitted with what was to be the standard armament of the He 162, twin 20-mm MK 151 cannons.

These ten pre-production models were subjected to the rigours of test flying by a number of different test pilots. With all the tests carried out the He 162 went into production and in order to keep track each aircraft was given an individual *werkenummer*, which indicated where it was built. He 162s that were built at the Marienehe facility were given the Werk-Nr 120 001, those built at the Hinterbrühl facility Werk-Nr.220 001 and those built at Junkers, Werk-Nr.300 001, and so on.

The first of the He 162s to come off the production line with the new numbering system were the He 162 V18 *(Werk-Nr.220 001)* and He 162 V19 *(Werk-Nr.220 002)* which showed

Front three-quarter view of the Heinkel He 162.

Instrument panel of the Heinkel He 162 showing its simplicity.

Underground factory for the construction of the Heinkel He 162.

that they had come from the Hinterbrühl factory. Both the aircraft were used for endurance trials. The V11 *(Werk-Nr.220 017)* and V12 *(Werk-Nr.220 018)* models were fitted with the Junkers Jumo 004D turbojet. These two aircraft were to be the prototypes for the He 162A-8. There was no V13 because of superstition, but the V14 *(Werk-Nr.220 015)* and V15 *(Werk-Nr.220 016)* models were retained for static test airframe development.

A new unit, the *Volksjäger-Erprobungskommando* under the command of *Oberstleutnant* Hans Bär, was set up at the end of January 1945, to evaluate the He 162. For the first three months the unit operated from Rechlin-Roggenthin, then was moved to München-Reim. At the same time ground crew training was being undertaken by *Fliegertechnischen-Schule 6* at Neuenmarkt and Weidenberg. The first unit to be supplied with the aircraft for combat was *Jagdgeschwader 1*, which was under the command of *Oberst* Herbert Ihlefeld. *I/JG 1* was ordered to hand over all their Focke-Wulf 190s to *II/JG 1* and carry out a conversion programme to the He 162 under the control of the factory test pilots. The *Gruppe* was reduced from four *Staffeln* to three (the personnel of the fourth *Staffeln* being absorbed into the other three), and the nine-week training began.

The need for a trainer was becoming a priority so *He 162 V16 (Werk-Nr.220 019)* and V17 *(Werk-Nr.220 020)* were converted into He 162S training gliders. The turbojet was removed and a second cockpit fitted behind the pilot with just basic controls. The wingspan was increased by almost 4 feet to 26ft. 10½ins., the vertical tail surfaces were enlarged and the undercarriage fixed in the down position. On the face of it, it would appear that almost all the aircraft that

Captured Heinkel He 162 about to be evaluated.

were coming off the production line were being used for testing purposes, but in fact a number were being put out for flight acceptance tests before being delivered to the Luftwaffe.

A new and less expensive undercarriage was designed and fitted to He 162 V20 (*Werk-Nr.220 003*) and initial tests of the undercarriage were carried out on 10 February 1945. Aircraft were taken off the production line and used for a variety of test purposes, for example He 162 V21, V22, V23 and V24 were all fitted with various designs of wing fillets and other modifications. The next two airframes, V25 and V26 (Werk-Nr.220 008 and 009) were prototypes for the He 162A-6, which had a slightly longer fuselage, and two short-barrelled 30-mm cannon. Flight tests were carried out on both these aircraft on 17 February 1945. Over the next two months armament trials and tests were carried out on the next four models V27, V28, V29 and V30 (*Werk-Nr.220 010–013*). Although a number of tests were carried out using the twin MG 151 of Mk.8 cannon, a large number of trials were carried out using a device called the *15er Wabe*. This was a honeycomb consisting of 15 tubes each of which contained a single R4M rocket missile. Two of these honeycombs were mounted beneath each wing and were situated 19 inches apart. The firing sequence was timed at one R4M rocket being fired every 70 milliseconds. In addition to this weapon a variation on the SG 117 machine gun called the *Rohrblocktrommel*, was tested at the same time. It consisted of seven 30-mm MK.108 barrels, each with automatic triggering, and containing seven shells in each barrel. Neither weapon was tested, as the war came to an end before the testing procedures began.

The appearance of the He 162A-9 powered by the BMW 003 engine combined a liquid fuel rocket with a turbojet. This was the prototype and if successful was to be manufactured as the He 162E-1. The He 162A-1, which had been used for testing a short-barrelled, low-velocity MK 108 30-mm cannon, had preceded this. During tests it was discovered that the firing of the gun created excessive vibration throughout the aircraft with the result that the forward

Excellent head-on shot of the Heinkel He 162 showing its simple, clean lines.

Hans Kuenneke standing beside his operational He 162A.

section of the fuselage would have to be re-stressed. As an interim measure the aircraft was fitted with 20-mm MG 151s, which were almost half the weight of the 30-mm cannons. The end result was that a 140lb trim weight had to be mounted in the nose of the aircraft so that the c.g. (centre of gravity) could be maintained. The aircraft was re-designated the He 162A-2. A number of other minor alterations were made to the He 162A-2, the wing area was marginally increased, the result of larger ailerons being fitted, and the span of the tailplane was increased from 8ft.8½ins to 11ft.1½ins. The BMW 003A was to have been installed in the aircraft but it wasn't ready so the Junkers Jumo 004D was installed.

Models He 162A-3 to -7 were to have been used as test beds but did not come to fruition and were never used. The He 162A-8 did, although it never went into production. A new Heinkel engine appeared, the Heinkel-Hirth 011A, also known as the HeS 11A. This had been destined to be the engine for the He 162B-1. The fuselage had been lengthened to 31ft. 2½in, which allowed for an additional 106 gallons to be carried and both the wing span and wing area were increased – the wing span to 24ft. 11½in. and the wing area to 129.167 sq.ft.

A twin engined version of the He 162B was studied using two Argus-Rohr pulse-jets. It was quite obvious right from the start that these engines were unsuitable for fighter aircraft, but as the war worsened for the Germans, anything and everything suddenly became worthy of consideration. An air of desperation was creeping into the Luftwaffe hierarchy. One of the uses put forward was that of a manned interceptor missile, much like the Japanese Okha (Cherry Blossom), a piloted bomb that was dropped from a 'Betty' bomber. The Heinkel team dismissed this idea and put forward two proposals; the first, a single As 014 Argus-Rohr pulse-jet of 1,100lb thrust to be fitted, or secondly, As 014 Argus-Rohr pulse-jets of 734lb thrust each. Both would have to have a re-designed wing structure to take the 194-gallon fuel tanks. It was obvious that these aircraft would have to be either towed into the air before proceeding under their own power, or catapult launched. Ernst Heinkel showed no real interest in developing this aircraft and it never left the drawing board.

The design team continued to progress with the He 162 and developed a swept wing version, the He 162C. This combined the B-type fuselage with the Heinkel-Hirth 011A turbojet. The wings were swept back to an angle of 38 degrees and had an anhedral angle two-thirds along the 26ft.3in. span. The V-type tail section was one that had been tested on the He 280 V8. The design team took it one step further and produced the He 162D, where the wings were swept forward. This necessitated moving the wing roots back along the fuselage where the trailing edge of the wing was in a line with the rear of the engine. The wingspan was decreased by almost 3ft. although its overall length was increased to 32ft.1½in. The proposed armament for both the C and D models was a pair of 30-mm MK 103 cannons, a variation of the *Schräge Musik* that had been installed in the Messerschmitt Bf.110. It enabled the pilot to fly underneath his quarry and fire upward into the belly of the aircraft. It was never installed.

There was one other variant, the He 162E, which was a proposed version of the He 162A-9 model, only fitted with the BMW 003R. This was a BMW 300A turbojet producing 1,764lb static thrust, coupled with the BMW 718 bi-fuel rocket producing 2,700lb static thrust. It was decided to use the rocket for climbing to operational height, where the turbojet would take over. Maximum speeds of around 630 mph at sea level were anticipated after the aircraft had climbed to a height of 16,500ft. in 1 min. 57 secs. It would carry 100 gallons of fuel for the turbojet and 2,640lb of fuel for the rocket motor. The war ended before it could be put to the test.

Despite the dead ends and final defeat, Ernst Heinkel had made a major contribution to the world of jet- and rocket-propelled aircraft.

7

HENSCHEL HS 132A

It was the failure of the Junkers Ju87 as a dive-bomber when faced with high-speed opposition that prompted the Henschel Flugzeugwerke to put forward a proposal for a turbojet dive-bomber. Designed by Dipl.Ing Friedrich Nicolaus, the Hs 132A was revolutionary in design inasmuch as the pilot flew the aircraft from a prone position. Trials had earlier been carried out on the Berlin B 9 a twin-engined, low-wing experimental monoplane. Powered by two 105-hp Hirth HM 500 air-cooled engines, this aircraft had been designed and built by the Flugtechnische Fachgruppe of the Technische Hochschule in Berlin-Charlottenburg. It had been discovered during tests that a pilot sitting in the normal position would black-out between 4 and 5g, but if in the prone position he could sustain as much as 10 and even 11g. After extensive testing at the Erprobungsstelle Rechlin, the data

Head-on view of the Henschel Hs 132A.

Artist's impression of the Henschel Hs.132A in action.

Remains of a Henschel Hs.132 found after the war after being cannibalised to repair other aircraft.

received proved beyond doubt that the prone position enabled the pilot to withstand much higher g-forces than the normal flying position.

Armed with this data the Henschel *Flugzeugwerke* set to work constructing the Hs 132A. The basic design of the aircraft consisted of a mid-wing monoplane with twin inclined tail fins and rudders and a fully retractable undercarriage. The wing was constructed by means of two wooden spars covered with plywood, into which the undercarriage retracted. The fuselage was circular and of an all-metal construction, which tapered towards the tail section. The pilot's cockpit, which was extensively glazed, was in the nose of the aircraft and streamlined into the fuselage. On top of the fuselage, mounted just aft of the cockpit, was a

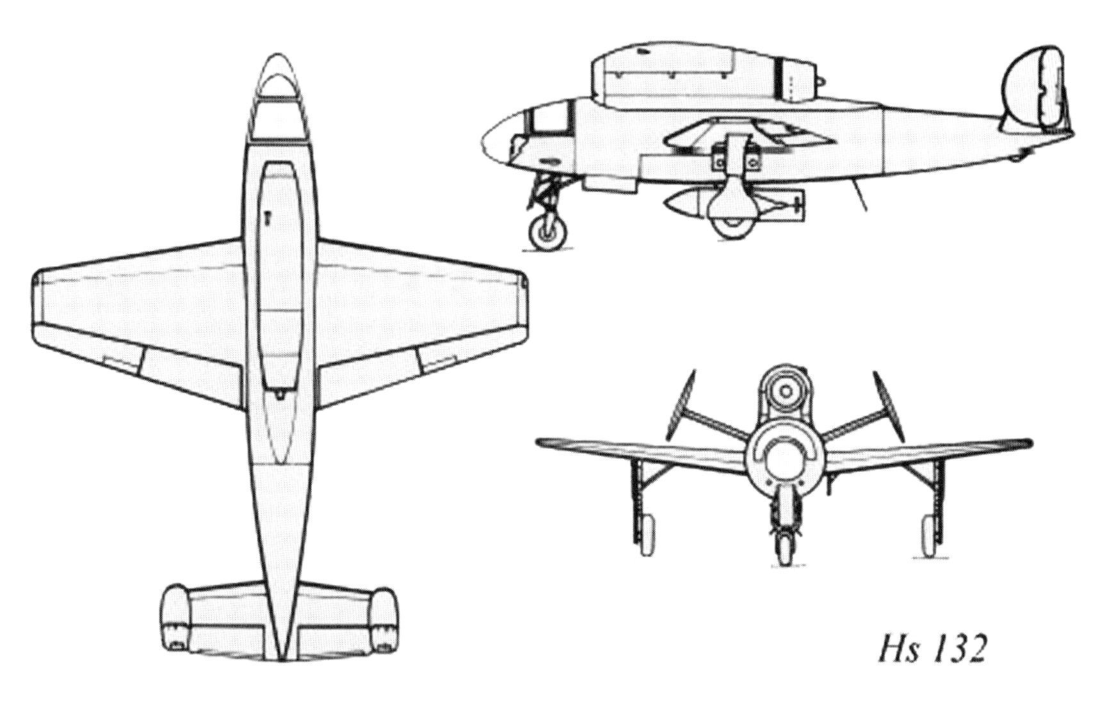

Three view drawing of the Henschel Hs 132A

Factory fresh Henschel Hs.132A.

single BMW 109-003E-2 turbojet of 1,760lb static thrust. The exhaust from the engine went between the two inclined vertical surfaces of the tail section.

The design of the aircraft was accepted by the *Technischen Amt* of the RLM, and resulted in six of the prototypes being ordered, two A models and four B models. The HS 132A V1 and

V2 were assigned as prototypes for a dive bomber fitted with the BMW 109-003E turbojet engine and with provision for a 500lb bomb which was to be semi-recessed into the fuselage.

The Hs 132B was to be used as a ground attack aircraft powered by a Jumo 004B-2 turbojet engine. Like the Hs 132A it would also carry a bomb, but in addition woud have two MF 151 20-mm cannons mounted in the nose beneath the pilot's prone position. The idea was that these aircraft could reach speeds in excess of 500 mph and, in the event of them diving to attack, anti-aircraft gunners would be unable to bring their guns to bear on such small, fast aircraft. It was further proposed that with the success of these aircraft and the availability of the Heinkel-Hirth 001A-1 turbojet, the development of the Hs 132C would commence.

The Hs 132C was to be fitted with the Heinkel-Hirth 011A-1 turbojet with 2,866lb static thrust. Armament was to be two MK 103 30-mm cannons, two MK 151 20-mm cannons and one 551lb or one 1,102lb bomb. There was a fourth model, the Hs 132D. This had an increased wingspan of 6ft.3in. making it 29ft. 10½in., which increased the wing area by 13 sq. ft. Early in 1945 construction of the Hs 132 V1 was well advanced and almost ready for flight test. Construction of the other models was also well under way but before production could begin the Russian forces advanced faster than had been anticipated, and the Germans were overrun.

8

ARADO AR 234

The Arado Ar 234 is regarded as the world's first pure jet bomber. During the final years of the war the Luftwaffe had based their hopes on this unique aircraft. The work to develop the Ar 234, or E370, as it was originally known, had started back in 1941 and initially it had been designed as a reconnaissance aircraft. Professor Walter Blume, one of the directors of the Arado Flugzeugwerke, and Ing. Rebeski had designed the aircraft as a high-winged monoplane powered by two Junkers Jumo 004 turbojets that were mounted in pods beneath the wings. The aircraft had no undercarriage and took off mounted on a three-wheeled trolley, which was jettisoned as the aircraft became airborne. Landing was carried out by lowering a retractable skid. The idea behind the E370 was that by losing the weight of the undercarriage the aircraft could carry additional fuel giving it a range of around 1,340 miles. This, together with an operating height of 35,700 feet at a speed of 485 mph, gave the aircraft an advantage over anything the Allies had, and a potential to carry out photo-reconnaissance flights anywhere in the West.

The proposals were put forward to the Luftwaffe who had hoped for an aircraft with an increased range, but they were satisfied with the remainder of the proposals and placed an order for two prototypes to be built, to be given the designation Arado Ar 234.

Within months the airframes were completed and just awaited the arrival of the Junkers Jumo 004 engines. They waited and waited but Junkers were having problems with the new turbojet engine. They were not the only one, BMW were also having their problems developing the turbojet. The Ar 234 airframes were put to one side whilst the company got on with other war work, then in February 1943 two Junkers Jumo 004 engines arrived. The excitement soon faded when it was discovered that they were only suitable for static and taxiing trials.

Then at the beginning of May 1943 a pair of flight-cleared Junkers Jumo 004 turbojet engines arrived, and by the end of July, after a series of ground tests, the aircraft was ready for its first flight. On 30 July 1943 with *Flugkapitän* Selle at the controls, the Arado Ar 234 took off from the airfield at Rheine near Münster. The flight itself was uneventful and all the tests carried out were successful, but the take-off trolley, when released, crashed to the ground when the retarding parachutes failed to deploy. The trolley was destroyed so another had to be built and despatched to the airfield. This of course held up the flight tests, but

Arado 234 V1 prototype on its launching trolley at the Brandenburg factory.

Arado Ar.234 lifting off its launching trolley on take off.

within weeks the aircraft lifted off again on its second test flight, and yet again the trolley was destroyed when the parachutes failed to open. It was decided that in future the trolley would be released the moment the aircraft became airborne, which in reality meant that the trolley barely left the ground. This system worked satisfactorily and was adopted thereafter.

Three more prototypes were ordered and by the end of September 1943 they too had been test-flown satisfactorily and interest in mass-producing the aircraft was growing daily. The Luftwaffe, dismayed at the way the war was going, started to look at the possibilities of using the Ar 234 as a bomber. It was discussed at the highest level, the RLM, who pointed out that because the aircraft took off on a trolley and taking into consideration its size, it would be extremely difficult to carry bombs under the fuselage. The alternative was to fit the aircraft with a retractable tricycle undercarriage. This was approved and two prototypes with undercarriage were ordered.

With the two prototypes ready for testing the programme appeared to be back on track, but then on 2 October 1943 *Flugkapitän* Selle was lost. He was carrying out a test flight of the second prototype when during the flight the port engine shut down. Typical of the professional test pilot that he was, Selle gave a running commentary over the radio so that notes could be taken on his predicament. Putting the aircraft into a glide, Selle noticed that there was excessive vibration coming from the elevators and the landing skid had become inoperative and would not extend. Lowering the landing skid by means of a manual operation, Selle asked if it had lowered because his instruments were not functioning. It was then realised that the port engine was on fire and it suddenly broke away from the wing and the aircraft plunged into the ground from 4,000 feet. The wreckage was examined and it was discovered that there had been a fire in the wing from the moment the engine had failed.

Tests continued on four of the trolley mounted Arado Ar 234s, Nos. 3, 4, 5 and 6. The sixth prototype was then fitted with four BMW 003 turbojet engines of 1,760lb thrust mounted

Extremely rare shot of an Arado Ar. 234 about to touch down on its landing skid.

in separate pods beneath the wings. The No.8 prototype was also fitted with four BMW 003 turbojet engines but these were paired in underwing pods. A third prototype aso using four engines had the provision of a Walter 109-500 rocket booster engine pod under the outer section of each wing to enable fully bomb-laden Ar 234s to take-off from short runways. The pods, each containing 1,100lb of additional thrust could run for 320 seconds before the fuel became exhausted. In the event of one of the pods not firing, a fail-safe system was in place that automatically shut down the other pod, avoiding an asymmetric thrust situation. The pods were then jettisoned and parachuted down to be recovered and refuelled to be used again, much like the Solid Rocket Boosters (SRB) used on the Space Shuttle fifty years later, those on a much larger scale.

The ninth prototype appeared in March 1944 and was known as the Arado Ar 234B and was the first model with a built-in undercarriage. Within weeks it had carried out its first test flight and the initial results were good enough to send the company at Alt Loennewitz in Saxony into a tooling-up production phase. Two Junkers Jumo 004B turbojet engines each producing 1,980lb thrust giving the aircraft a maximum speed of 461mph, powered the Ar 234B version. When carrying the maximum bomb load however, the speed was reduced by 45–50mph. This also reduced the maximum range of the aircraft and in short meant that carrying a 1,100lb bomb load the Ar 234 had a range of around 300 miles. When operating in a purely reconnaissance role and fitted with two 66 gallon auxiliary fuel tanks the range could be extended to 450 miles.

Bomber

The bomber version of the Ar 234B was designed to use three types of bombing attack, the first being the 'shallow' bombing technique which saw the aircraft nose over at 16,000

Four-engined version of the Arado Ar.234 being manhandled into position for take-off.

Arado Ar.234 preparing to take off on a test flight with dummy outboard engines.

feet then level off at 4,500 feet. The pilot used a periscope, which protruded from the top of his cockpit canopy, to sight the bombs. Then there was the low altitude attack method that was only used during poor visibility. The pilot simply ran in and dropped the bombs in line of sight. The high altitude method was without doubt the most interesting and brought into use a computer, which controlled the autopilot. When about 20 miles from the target, the pilot would engage the Patin three-axis autopilot, which released the control column enabling the pilot to move it to one side. Once this was done the pilot released his seat harness and moved forward to the bomb-aiming position and positioned himself over the Lotfe bombsight. A small, simple computer was automatically engaged that connected the bombsight to the autopilot. The pilot then moved the reticule (the grid of lines drawn over a map) in the bombsight over the target, which then fed a signal to the autopilot via the computer. This, in effect, meant that the bombsight flew the aircraft through its bombing run. On reaching the target the bombs were released automatically, the pilot then just leant back, tightened his harness, released the autopilot and turned for home. Not so remarkable today, but this was 1944 and to have such an advanced system as this demonstrated the progress that Germany had made.

This wasn't the only new innovation that the Arado Ar 234B possessed. It was also fitted with the world's first braking parachute that had been designed to shorten the landing run of the aircraft.

Reconnaissance

The first of the 20 Ar 234B production models ordered arrived at the beginning of June 1944, less than one year after the appearance of the Ar 234A. Two of the Ar 234A prototypes were fitted with cameras and assigned to 1st *Staffel Versuchstaffel der*

Oberkommando der Luftwaffe (Luftwaffe High Command Trials Detachment). This was a special reconnaissance unit headed up by *Oberleutnant* Horst Goetz who started training with the Arado Ar 234A. This model still had no undercarriage and took off attached to the tricycle trolley. Goetz had a couple of scares when the trolley failed to release on take-off, but Goetz was a highly skilled pilot and had no trouble landing the aircraft still attached to the trolley.

The cameras, Rb 50/30s with 50cm long-focus lenses, were mounted in the rear section of the fuselage in tandem and placed in a downward inverted 'V' position 12 degrees apart, which gave a spread of six miles whilst photographing from 33,000 feet. It was around this time that the Allied forces came ashore at Normandy on D-Day. The Luftwaffe sent photo-reconnaissance units to gain urgently needed information, but they were met by scores of fighters that inflicted severe losses on them. Horst Goetz's unit was ordered to get ready and was moved to Juvincourt near Reims in preparation for missions over the advancing Allied troops. Both Ar 234A aircraft took off from their base at Oranienburg but Goetz's aircraft had engine trouble so he had to abort and return to base. His fellow pilot in the other Ar 234A, *Leutnant* Erich Sommer continued on and landed at their new base at Juvincourt. On landing there the aircraft was hoisted on to the back of a low loader and taken to a hangar on the other side of the airfield, where it stayed until its launching trolley arrived two weeks later by train from Oranienburg. With the trolley came all the spares and equipment necessary to keep the two aircraft operational.

On 2 August 1944 the world's first jet photo-reconnaissance mission took place, when Erich Sommer opened the throttles on his Arado Ar 234A and lifted into the air aided by two booster rockets. Fifteen seconds after lift-off he released the spent boosters and climbed the aircraft to a height of 34,000 feet over the advancing Allied forces. Without a cloud in the sky Sommer opened the camera doors in the fuselage, tripped the camera switch and dropped to 1,700 feet at a speed of 460mph. Keeping the aircraft straight and level he started the cameras rolling, the shutters opening and shutting every 11 seconds. Each run lasted about ten minutes and after the third run Sommer shut the cameras off and headed back to Juvincourt. Extending the landing skid, he touched the aircraft down onto the runway and slid to a halt. He had barely come to a stop before people were swarming all over the aircraft and the cameras were removed. They were on their way to be developed long before he himself had got out of the cockpit, such was the urgency. Erich Sommer and his Arado Ar 234A had accomplished more in 30 minutes than the whole of the Luftwaffe reconnaissance units had in the previous two months.

The results from the photographs, which took twelve photographic analysts two days working flat out to interpret, were to stagger the German High Command. The Allies had landed almost 2 million men, over a million tons of supplies and almost a quarter of a million vehicles of all shapes and sizes. A fully detailed examination of the photographs took over two weeks but by this time the Allies had moved inland rapidly, making the photographs redundant. *Oberleutnant* Horst Goetz finally arrived with his aircraft on 3 August, and during the next two weeks the two Arado Ar 234As carried out a total of 13 photo-reconnaissance missions, keeping the army field commanders up-to-date with photographic information on the progress of the Allies.

As the American tanks bulldozed their way toward Reims, Goetz was ordered to move both the aircraft and his entire unit from Juvincourt to Chievres. Up to this point the Allies were not aware of the existence of these two fast, high-flying reconnaissance jets that were photographing their progress and because of this they were never fired upon, either from the ground or in the air. That was until Horst Goetz circled the airfield at Chievres and was about to enter the landing pattern. The anti-aircraft crews on the airfield, nervous to say the least, had never seen anything like the Arado Ar 234A and opened fire. For once they were accurate and one of the shells hit Goetz's aircraft just behind the cockpit knocking

Arado Ar.234 coming into land. Note the retractable undercarriage.

out all the electric and hydraulic systems. Realising he couldn't land at Chievres he headed back to Juvincourt where he made an emergency textbook belly-landing sustaining very little damage. Unfortunately, he had failed to notice a Messerschmitt Bf 109 also on the runway taking off. The fighter ran smack into the rear of Goetz's aircraft severing the tail and causing damage to the extent that the aircraft was written off. Horst Goetz suffered minor injuries and was back flying within the week.

Leutnant Erich Sommer managed to land his aircraft at Chievres but within days was moved on to Volkel in Holland. Two days later the airfield was subjected to a heavy bomber raid by 100 RAF Lancaster bombers and, although Sommer's aircraft escaped damage, the airfield itself was badly damaged. Normal operations from the airfield were suspended until some of the craters that pockmarked the airfield were filled. Within days the Ar 234A was on the move again, this time to Rheine near Osnabrück where, in effect, all operational sorties by the Ar 234A ceased as the Ar 234B, with its retractable undercarriage, was becoming more available.

A new unit under the command of *Oberleutnant* Horst Goetz was formed at Rheine, the *Kommando Sperling*. The unit was equipped with nine Arado Ar 234Bs and their sole role was as photo-reconnaissance aircraft. Although the aircraft could outrun anything the Allies had to offer, the Ar 234b was particularly vulnerable during landings and take-off. Horst Goetz decided that his unit would only fly when there was no danger of there being enemy aircraft in the vicinity. This was becoming increasingly difficult because of the rapid advances being made by the Allied forces. Surrounding the airfield with anti-aircraft guns and making pilots stay on full power when they neared the airfield until told it was safe to land was one way of dealing with the problem.

The unit became operational almost immediately on being formed and it started to carry out photo-reconnaissance missions over France, Belgium, Holland and England. A two-hour flight over England became the norm and only occasionally was there any contact with Allied fighters and that was very brief. It was not until 21 November 1944 that the first report of the existence of the Arado Ar 234 became known to the Allies. This came about when P-51 Mustangs of the 339th Fighter Group were escorting B-17 bombers over Holland when the aircraft was spotted flying about 1,000 feet above the bomber formation. Before the P-51 Mustangs could engage the aircraft it accelerated away.

From the beginning of December 1944, as more and more Arado Ar 234Bs came off the production line, *III Gruppe* of *Kampfgeschwarder 76* had been working-up with the aircraft at Burg near Magdeburg. A number of trials were carried out towing a converted flying bomb. Stability problems caused the project to be cancelled.

To the Ardennes

On 17 December Hauptmann Diether Lukesch, Commander of *9 Staffel*, was given orders to move 16 Ar 234Bs to Münster-Handorf in support of the German offensive in the Ardennes. Within four days all the aircraft and ground crews were at their new base and being prepared for action. Then the weather turned and there was no flying until Christmas Eve,

Arado Ar.234C with twin Jumo engines.

when at 1014 hours, the first of the bombers took off. Diether Lukesch was the first off and he was soon joined by the remaining eight Ar 234Bs. Their targets were the railway yards and factories in the city of Liège. Each aircraft carried a single 1,100lb bomb so there was no room for error. The bombers reached their objective and one by one selected their targets and dropped their bomb. Joining up afterwards they headed for home, reaching their base at Münster-Handorf safely. That afternoon a second raid was carried out on the city and the same targets were attacked.

On Christmas Day there was no let-up and the nine Ar 234Bs once again took to the air to bomb the city of Liege. This time however they were met by Hawker Tempests from No.80 Squadron, Royal Air Force. As the Ar 234Bs scattered, one, flown by *Leutnant* Alfred Frank, was hit by gunfire from one of the Tempests flown by Pilot Officer Verran. Frank managed to limp away after the Tempest had run out of ammunition but the damage was done and he had to crash-land the aircraft in Holland. A second Ar 234B took a hit in the engine during the fight with the Tempests but the pilot managed to make it back to base before he too crash-landed, seriously damaging the aircraft. The remaining seven aircraft made it back to their airfield unscathed.

On 1st January 1945 the Arado Ar 234Bs were in action again, this time they took off in the early hours of the morning on a circular route via Rotterdam, Antwerp, Brussels, Liège, and Cologne and then back to their base. The intention was to carry out a reconnaissance mission on the weather over Holland and Belgium in advance of Operation *Bodenplatte*, a massed attack on Allied airfields. To conceal the real reason for the mission, they dropped their bombs on Brussels and Liège. On their return to Münster-Handorf, six Ar 234Bs of *9 Staffel*, each armed with a 500lb container carrying twenty-five 33lb anti-personnel bombs, took off to carry out a raid on the Allied airfield at Gilze Rijen in Holland. This was almost the last raid to be carried out in January as the weather worsened to the extent that there was almost no flying from either side.

By the middle of January 1945 only 17 Arado Ar 234Bs were in operational service, although *I Gruppe* and *III Gruppe Kampfgeschwader 76*, under the command of Major Hans-Georg Baetcher, were re-equipping with the Ar 234B. Of the 148 Arado Ar 234Bs ordered by the Luftwaffe in the middle of 1944 only a very small proportion of the aircraft had

Arado Ar.234 taking off with the assistance of two liquid fuel rockets.

been delivered, and of these only a small number were in a state of operational readiness. Incessant attacks on the supply trains and roads were preventing spares and fuel getting to the airfields.

At the end of January the whole *Gruppe* had been converted to Ar 234Bs. Within days two of the *Staffeln*, Nos 7 and 8 with 18 of their aircraft, were sent to their new operational base at Achmer. As they arrived over the airfield and started their approach, they were 'bounced' by Spitfires from No.401 Squadron (Canadian). The bombers attempted to scatter, but the Spitfires were in amongst them resulting in three Ar 234s being shot down, two badly damaged aircraft and the loss of two pilots.

Refitting their *Staffeln*, the *Gruppe* carried out a series of very successful raids during February. Sixteen of the bombers attacked British troops near Cleve inflicting heavy casualties and on 21 February 37 sorties were flown against British troops near Aachen, again inflicting very heavy casualties. Moves were made to convert *II Grup*pe with the Ar 234B and by the end of February *6 Staffel* had been completely re-equipped and was operational.

The Bridge at Remagen

On 7 March came one of the severest setbacks for the retreating German Army. As they retreated the Germans were destroying the bridges that spanned the Rhine in an effort to slow down the advance of the Allies. The Ludendorff Bridge at Remagen was one such

Arado Ar.234C prototype with twin Jumo engines.

Arado Ar.234 taking off leaving a trail of black smoke.

bridge, only this time the demolition charges set to blow the bridge did not destroy it. Badly damaged, the bridge was still usable and after a brief battle with inferior local troops the Americans took the bridge. Troops and armour began to pour into Germany night and day, so Göring decided that the bridge had to be bombed and assigned bombers from *III Gruppe* the task. The first attack on the bridge was on 9 March when three Ar 234Bs carried out a low-level attack on the bridge, but the Americans had been expecting an attack and had set up anti-aircraft guns around the bridge. One of the bombers was shot down and the bridge,

Major Hans-Georg Blucher about to get into the cockpit of his Arado Ar.234.

although hit, remained intact. Two days later another attack resulted in a little damage but nothing significant.

Three days later, with heavy cloud obscuring the bridge, the Germans tried a new approach and carried out a total of 18 attacks but each time by a single aircraft and horizontally from a height of 25,000 feet. These were unsuccessful, as were the ones the following day. By the 14th the sky had cleared and the cloud lifted sufficiently for a clear run at the bridge. Eleven Ar 234B bombers took off and headed for the target. They were met by large numbers of British and American fighters who had started flying protection patrols around the area. In the ensuing melee four of the bombers were shot down but no damage was caused to the bridge. All the time troops and armour were flooding into Germany over the bridge and the surrounding area on the German side was securely in American hands. The American engineers had also built pontoon bridges alongside the Ludendorff Bridge when they realised that the accumulative damage of the German engineers and the bombing had weakened the bridge to a point where it was considered dangerous. The Ludendorff Bridge finally succumbed to the damage and collapsed into the Rhine on 17 March.

First Loss Before the End

The *Gruppen* continued their attacks on Allied positions and in one attack, led by *Leutnant* Croissant, the Allied airfield at Melsbroek was bombed and damage was caused to one of the

Arado Ar.234C on its launching trolley.

Arado Ar.234s on the flight line of unknown Jagdegeschwader.

RAF's operational jet fighters, the Gloster Meteor of No.616 Squadron, RAF. Two other RAF jet fighters were scrambled but the Ar 234Bs were long gone. Had they caught up with them it would have resulted in the first jet-to-jet aerial combat.

The Arado Ar 234B, which had been operating as a photo-reconnaissance aircraft since its introduction into the Luftwaffe, had its first casualty in February 1945, when it was intercepted whilst on a mission and shot down by a Hawker Tempest of No.274 Squadron, RAF. The Ar 234B, flown by *Hauptmann* Hans Felden, was about to land at Rheine when Squadron Leader David Fairbanks, who had been chasing it, raked it with machine gun fire. The aircraft crashed on the airfield killing the pilot immediately.

Whilst the Arado Ar 234B continued to carry out photo-reconnaissance missions and bomber sorties, the Arado factory produced a four-engined version – the Ar 234C. Powered by four 1,760-pound thrust BMW 003 engines, the first test flight was almost its last. The test pilot Peter Kappus had just taken off when he noticed a sudden increase in engine noise and cut engine power to the No.2 port engine as the rpm was running at 11,000 instead of the maximum of 9,600. Deciding to take the aircraft back to the field, he put the aircraft into a turn and increased the power on the other engines. Unknown to him a sheet of flame from the crippled engine had spurted out almost the length of his aircraft. Upon landing he was immediately surrounded by fire crews who managed to put the fire out before it could envelop the aircraft.

This was an ongoing problem with the BMW 003 engine, which, although thought to have been rectified, never had been. Fourteen of the Ar 234Cs were built before the whole Arado plant was blown up to prevent it falling into the hands of the advancing Soviet Army. The days of the Luftwaffe were numbered, through lack of fuel, ammunition, bombs and spares and of course the overwhelming air forces of the Allies. The Arado Ar 234, despite all its potential, never really achieved a great deal as a bomber – but as a reconnaissance aircraft it really excelled.

9

JUNKERS JU 287

In 1943 with the war slowly turning in favour of the Allies, the Luftwaffe appointed Oberst Siegfried Kneymeyer to the post of Chief of Technical Air Armament. Kneymeyer was both a pilot and a scientist and had carried out the first high-altitude photo-reconnaissance mission over North Africa in a Junker Ju 86P-2 earlier in the war. No stranger to controversy, Kneymeyer suggested that to enable the Luftwaffe to compete with the numerical superiority of the Allied aircraft, they needed a 'cushion' of speed superiority of around 93 mph. The Me 262 jet fighter had borne out the theory to some extent, but Kneymeyer's intention was to have bombers with this margin of superior speed. The Arado

Hugo Junkers (3 February 1859–3 February 1935).

Junkers Ju.287 showing the four engines – two either side of the fuselage at the nose and two underneath the wings.

Junkers JU 287 VI being prepared for a test flight.

Junkers Ju.287 under repair.

Company had produced the Ar 234 but this was only slightly larger than the Me 262 and could only carry one bomb.

On hearing of a bomber project being worked on at the Junkers plant by Dipl.Ing. Hans Wocke, Siegfried Kneymeyer put forward his ideas. The resulting project looked at a turbojet powered bomber with a wing sweep of 25°, which would allow a heavy bomber to exceed Mach 0.8 in level fight, although the sweep back of the wing would cause stability problems at some low speeds. With trials in a wind tunnel using varying wing sweeps and angles, Wocke came up with the suggestion that the aircraft would benefit not only from a greater sweep angle, but also with the wings being swept forward rather than back. This radical thinking was not new; Professor A.Busemann back in 1935 had proposed a swept forward wing design at a conference in Rome and trials had been included in research programmes by the *Deutches Versuchsanstalt für Luftfahrt* – DVL (German Aviation Experimental Establishment).

The problem of stability was still cause for concern, but it was thought that the forward sweep of the wing would only cause instability problems at high speed, where it could be controlled more easily. The RLM were convinced that this proposal merited the building of a prototype and awarded Junker a contract for the one aircraft, to be designated the Ju 287.

The design work was placed under the control of Dipl.Ing. Ernst Zindel and as work on the construction of the aircraft neared, it was decided to carry out a flight test on the radical new wing that had been proposed. To do this the fuselage from Heinkel He 177A was fitted with a forward swept wing, whilst the tail section of the aircraft used Ju 388 parts. The undercarriage, which was fixed for the trials, consisted of main wheels from a Ju 352 and twin nosewheels from a crashed Consolidated B-24 Liberator, all of which had covers on them. The aircraft was powered by four Junkers Jumo 109-004B-1 turbojets each producing 1,980lb static thrust. Two of the engines were mounted either side of the forward section of the fuselage and one beneath each of the trailing edge of the wings. It was decided that although they thought there was sufficient power with the four engines, it would aid the take-off if each engine had a Walter 109-501 jettisonable rocket unit fitted beneath it. This gave the engines an additional 2,200lb of static thrust. To aid braking a brake parachute was installed in the tail.

On 16 August 1944 with *Flugkapitän* Siegfried Holzbauer at the controls, the aircraft, designated the Ju 287 V1, rocketed off the runway at Brandis in a cloud of black smoke and with a roar that could be heard for miles around. It carried out a number of successful low-speed tests to evaluate the wing. Interestingly enough, as the Ju 287 was carrying out its first flight tests from Brandis, the first Me 163B Komet rocket fighters were making their operational debut on the same airfield.

The fuselage and wings were covered in small tufts of wool, and a camera (mounted at the rear of the fuselage and just in front of the tailplane), recorded the movements of the tufts. Problems were discovered when the aircraft was put into tight turns at high speed. The wings flexed at this point and extra loading on the elevators had to be provided to counteract the force. There were other relatively minor problems but to further stiffen the wings structurally was not an option, so it was decided to mount the engines forward of the wing's main spar to balance it. This arrangement was to be fitted into the Ju 287 V2 where two BMW 109-003A-1 turbojet engines were to be mounted beneath each wing and one either side of the forward section of the fuselage.

OKB-1EF 140, the Russian version of the Junkers Ju 287.

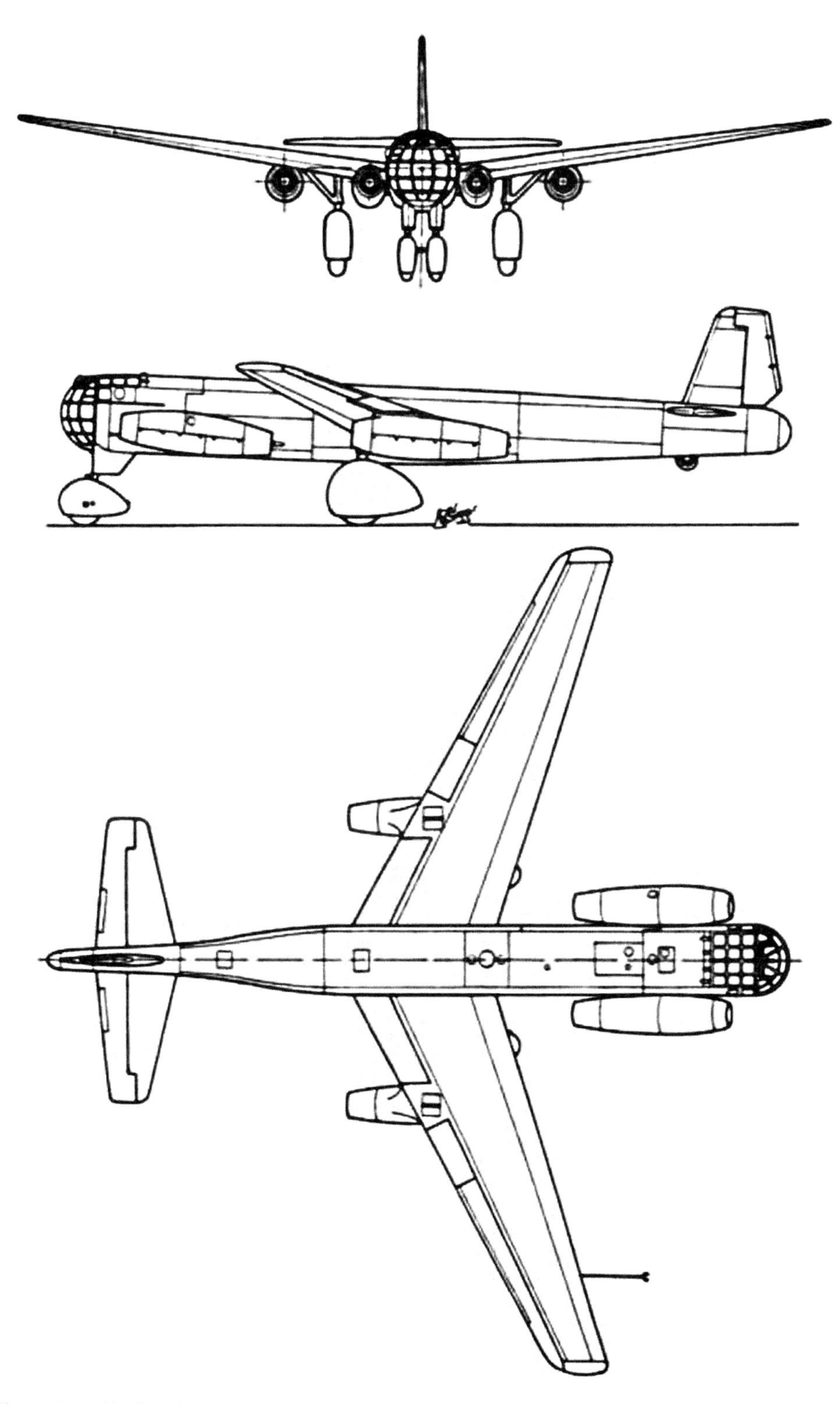

Three-view of Junkers Ju 287.

Junkers Ju 282 side view.

The only things that were the same as the V1 prototype were the wings. A newly designed fuselage had a larger nose and an increased glazed area whilst the wheels of the undercarriage retracted into the fuselage, but the oleo legs, attached to the main wing spar, retracted into the wings.

Fighters, not Bombers

In July 1944 the Luftwaffe High Command decided that it was fighters not bombers that were required because Germany was now on the defensive, so the project was shelved. Junkers continued to construct the Ju 287 V2, albeit at a very slow pace. For almost a year the project lay forgotten until March 1945 when for some unknown reason an order came out of the blue from the RLM to put the Junkers Ju 287 into production. Fortunately, the company had still kept their finger on the pulse of the project and were able to tool up the production lines; but like a number of other turbojet projects it was the engines that were to let them down. Problem after problem meant that the engines were not ready before the war ended, and although a number of the aircraft were built, not one saw combat.

But this was not the end of the Junkers Ju 287 project. Captured by the Russians, the engineers and all the plans for the aircraft were taken back to Russia and the aircraft constructed. After tests flown by German test pilot *Flugkaptitän* Dülgen, the aircraft was rebuilt, but this time with the wings in a swept back configuration. Reports state that it was a very satisfactory aircraft and furnished the Russians with a great deal of information on jet-powered flight.

10

MESSERSCHMITT ME 262

In 1938 the RLM issued a requirement for a two-engined jet fighter aircraft given the designation P.1065. A number of aircraft manufacturers were working on projects with Ernst Heinkel well in the lead, but in Augsburg, Messerschmitt were just completing the airframe for their P.1065 V1 model. Designated the Me 262 by the RLM the project was already being slowed down by the continuing problems with the BMW 003 engine, which had been earmarked for the Me 262. Bench tests had been carried out on the engine and a very disappointing 570lb static thrust was all the engine could produce. The design requirement was for an engine that produced 1,500lb static thrust and this was nowhere near. Messerschmitt turned to Hugo Junkers who, together with Dr. Anselm Franz, had been

Prototype Me 262 V1 fitted with a Jumo 210G piston engine.

working on a parallel engine since 1939. Despite all the promises and hopes it turned out to be no better than the BMW 003. Since the airframe was now complete and the only thing preventing its test was a powerplant, Messerschmitt suggested to the RLM that they could install two Walter HWK R II-203b rocket motors in the airframe. Although the flying time of the Me 262s would only be measured in minutes, it would give them an evaluation of the aircraft's aerodynamics.

Approval was given and work started to install the engines. There were rumblings amongst the engineers at Messerschmitt who were a bit reluctant to try this method, but they did not want to create friction between themselves and the RLM. Official backing from this quarter

Prototype Me 262 taking off on a test flight.

Sixth model of the Me 262. This was the first to be fitted with a retractable nose wheel.

Me 262-1a night fighter with radar antenna mounted on the nose.

Messerschmitt Me 262 front three-quarter view.

Me 262B-1a trainer.

Me 262 taking off towing a Deichselschlepp (wheeled bomb).

Test pilot Karl Bauer in the cockput of the prototype Me 262.

could bring a great deal of financial support for other projects, so it was decided to go along with the suggestion. One of the main reasons for the reluctance to use the rocket motors was that the fuel they used, the mixture of C-Stoff and Z-Stoff, was inherently unstable. The engineers at Messerschmitt delayed the airframe's readiness with the idea that the BMW 003's problems might be sorted before the installation of the Walter R II-203b rocket

Messerschmitt Me 262 PC+UC V3.

motors, but it was not to be. Then the *Technische Amt* of the RLM stepped in with concerns that the use of the rocket motor and its unstable fuel was an unacceptable risk. After talks with Messerschmitt's engineers, permission to install the rocket motors was withdrawn, which left Messerschmitt looking for an alternative power source. The 730-hp Junkers Jumo 210G 12-cylinder, liquid-cooled, inverted **V**-engine was chosen to test the airframe. The engine was installed in the nose of the aircraft and on 17 April 1941 the taxi trials of the Me 262 V1 (PC+UA) with test pilot Fritz Wendel at the controls took place at Augsburg. Late the following day the aircraft took to the air on its initial test flight just before the sun disappeared over the horizon. The reason for this was to keep the secret aircraft away from prying eyes. The first flight of the Me 262 V1 lasted just 18 minutes.

To say the aircraft 'took to the air' was over-optimistic, it *struggled* into the air and although there was very little testing of systems, its basic handling and aerodynamics were good. The second test flight lasted 31 minutes and thereafter the average length of a test flight was 30 minutes. During the following year the aircraft underwent testing by various test pilots from the Rechlin centre. Whilst the testing went on, the second Me 262 V2 was being constructed using the information being gathered from the testing of the V1. BMW suddenly appeared with their improved 003 engines, but the Messerschmitt engineers were wary and decided to put the engines into the V1, but leave the Junkers Jumo piston engine in place as a safety factor. Because of lack of faith in the BMW engine, the ground tests carried out using the engines were not very satisfactory. News filtered back to Messerschmitt that the Junkers 109-004 engine was progressing well and producing a static thrust of 1,320lb. After completing a ten-hour bench test on Christmas Eve without missing a beat, things looked good. Then, when the second engine was tested malfunctions manifested themselves. This of course put the programme back and it was the spring of 1942 before the engine could be flight-tested when it was fitted into a converted Bf 110C.

Head-on view of the Me 262 bomber version.

Engine Switch

On 1 February 1942 the first flight of the V1 with the two BMW 003 turbojet engines and the Junkers Jumo piston engine took place. Test pilot Fritz Wendel powered the two turbojet engines and the piston engine but the aircraft only just cleared the hedge at the end of the runway. Seconds later, the port turbojet engine flamed out followed seconds later by the starboard one. The Junkers piston engine struggled successfully to keep the aircraft in the air and Fritz Wendel managed to bring the aircraft round to make a safe landing. The engines were removed and it was discovered that the compressor blades in the engines had broken off, showing weaknesses in the metal used in their construction. It was to be over a year before the next generation of BMW turbojet engine appeared.

Quick to take advantage of BMW's problem, Junkers came forward with their Junkers Jumo 004 turbojet engine. Messerschmitt decided to go with the Jumo engines and made arrangements to fit them into the Me 262 V3. The RLM, concerned at the lack of progress, cut the contracts for the prototypes to just five airframes, but the 20 pre-production order was to stay, on the condition that there were successful demonstration flights using either the Junkers Jumo or BMW engines.

The engines arrived at the airfield at Leipheim (the runway was considerably longer than the one at Augsburg), but with them came a report expressing concerns about their reliability. It was decided to run additional bench tests then install them in a converted Messerschmitt Bf 110 for flight tests. All the while, work was going on to complete the construction of the Me 262 V3, which was to be the world's first jet fighter. After 50 hours of testing the engines were installed in the Me 262 V3 and the Luftwaffe awaited the first flight. But problems elsewhere were causing delays. Heinkel had run into trouble with the Heinkel-Hirth 001 turbojet engine. Just after taking off on a routine test flight, one of the turbine blades sheared off, causing the engine to catch fire. The test pilot Fritz Schafer managed to get the aircraft down for a wheels-up landing. Doubts were being cast over the reliability of these jet engines, first the BMW and now the Heinkel-Hirth. Coupled with all this was the news that one of Messerschmitt's staunchest supporters, *Generalfeldmarschall* Ernst Udet had committed suicide and his place had been taken by *Generalfeldmarschall* Erhard Milch. The problem was, as mentioned earlier, Erhard Milch and Willy Messerschmitt had an intense dislike for each other that went back over a number of years. This manifested itself when the RLM accused Messerschmitt of spending too much time and money in developing rocket and jet engine projects and not enough on the conventional aircraft the Luftwaffe desperately needed. One of these aircraft was the Me 210, which had been earmarked as a replacement for the Bf 110. The design was full of innovative ideas but the aircraft itself had so many handling problems that it became a nightmare to fly. One test pilot, Fritz Wendel, had to bail out after the tailplane broke off.

The *Generalluftzeugmeisteramt* (Director General of Luftwaffe Equipment), unaware of these problems, had ordered that the production lines for the Bf 110 be phased out and the factories be tooled up in readiness for the Me 210. The aircraft was declared totally unstable and the production of the Bf 110 restarted. It was estimated that the delay cost the Luftwaffe 600 aircraft in production losses, a huge shortfall laid squarely at the feet of Willy Messerschmitt by Erhard Milch.

Me 262A-1a of Kommando Nowotny being prepared for a mission.

The Me 262 V1 and V2 were still carrying out trials with the BMW 003 engines that only further demonstrated their unreliability. Erhard Milch called the Luftwaffe's Chief of Engineering, Dipl.Ing. Lucht, to a meeting where Lucht complained bitterly about the problems he was experiencing with the Messerschmitt factory. Milch saw this as an opportunity to get rid of Willy Messerschmitt and contacted Messerschmitt's deputy Franz Seiler and told him of his intentions. Seiler immediately told Messerschmitt what was going on and a board meeting was hurriedly convened. At a highly charged meeting it was decided the only way out of this was to get Milch to change his mind, so it was decided to demonstrate the Me 262 V3 and the top secret Me 309 to some of the top Luftwaffe officials.

On 18 July 1942, with test pilot Fritz Wendel at the controls, the Me 262 powered by two Junkers Jumo 004A-0 turbojet engines lifted into the air and into the history books – the world's first jet aircraft had arrived. After carrying out a series of turns and manoeuvres lasting just 12 minutes, Fritz Wendel brought the aircraft back down for a perfect landing. A second flight later that morning was equally successful. The following day Lucht arrived to inspect the company and to tell Willy Messerschmitt he would have to stand down as chairman of the company or it would be turned over to a government-appointed official. Incensed, Messerschmitt had no alternative but to relinquish control to Seiler, but at least this was someone who thought the same as he did. Messerschmitt did not leave the company but retired to the design office where he continued to work.

A number of tests were carried out over the next two months but there was the continuing problem of the engines. It was also decided to redesign the undercarriage giving it a tricycle configuration. But during the sixth flight test on 11 August 1942, one of the test pilots from Rechlin, Heinrich Beauvais, attempted to take off in the Me 262 V3. Fritz Wendel, who himself had had problems getting the aircraft off the ground, assisted Beauvais by positioning himself alongside the runway at the point where Beauvais should gently touch the brakes so that the tail raised, then pull back on the stick and lift off. After two attempts when Beauvais had to shut down and abort the take-off, a third attempt was undertaken. Opening up the engines he hurtled down the runway, but the aircraft would only just unstick and he careered through a cornfield at the end of the runway. The wingtip dug into the ground flipping the aircraft on to its back causing extensive damage. Beauvais escaped with only a few cuts and bruises.

Trials continued on the V1 whilst the V2 was fitted with the Jumo 004A-0 engines. Then on 1 October 1942 the V3 was ready for its first test flight flown by Fritz Wendel. Heinrich Beauvais carried out the second test flight of the V3, on the same day, this time without incident. These were the last test flights for the year but two more prototypes, V4 and V5, were in the process of being built. The RLM were now impressed with the aircraft and ordered a further five prototypes (V6–V10) and ten pre-production models, all of which were given V prefixes.

The end of 1942 saw Germany on the receiving end of defeats; Rommel had been pushed back from El Alamein and the German Army in the East had suffered serious setbacks at the hands of the Russian Army. The Americans had now entered the war and were pouring in equipment to bolster the Allied cause. It was understood in some quarters that once the

Me 262A-1a with a 50mm cannon mounted in the nose. The trials were unsuccessful.

American aircraft factories got into full production, aerial superiority would very quickly swing away from Germany. The RLM started to demand greater output from the aircraft manufacturers but this was unrealistic. They also created a priority list of jet aircraft known as the 'Vulcan Projekt'.

At the beginning of 1943 the Me 262 V1 emerged with a new wing shape and the removal of the Jumo 210G piston engine from its nose. Two Jumo 004A engines had been fitted and it was also fitted with a pressurised cockpit and three 15 mm MG 151 Mauser cannons in the nose. It was decided that the production model Me 262 would carry a 1,100lb bomb, six 30 mm Mk 108 cannons, or four Mk 108 cannons plus two 20 mm MG 151/20 Mauser cannons.

The time had come for the Me 262 to be evaluated by the Luftwaffe combat pilots and *Hauptmann* Wolfgang Späte, commander of *Erprobungskommando 16* (Test Unit 16) was assigned the task. He had evaluated the Me 163-rocket fighter and had considerable experience of both combat and testing procedures. His first flight, on 17 April 1943, was in the Me 262 V3 and he was immediately taken by its ease of handling. During a second test flight he carried out a steep banking turn at 9,840 feet, throttling back the engines as he did so. The move upset the airflow to the engines causing them to flame-out. Putting the aircraft into a shallow dive he attempted to restart the engines, managing to start the starboard engine at 4,900 feet and the port engine at just 1,500 feet. Bringing the aircraft safely back he was very impressed despite the problem of the flame-outs. Flying was suspended for the rest of the day so that the engines could be examined for any damage but none was found and the aircraft was declared fit to fly again. At the same time the Me 262 V4 (PC+UD), the fourth prototype with two Jumo 004A engines, had been finished and was ready for testing. It was to be 15 May 1943 before it finally arrived at Lechfeld to join the testing programme and it was also the last prototype to be fitted with the conventional landing gear.

Galland's Seal of Approval

Hauptmann Späte had reported back to *General der Jagdflieger* Adolf Galland on the results from the testing of the Me 262. Adolf Galland had known about the programme almost from its conception and had kept tabs on its progress. He approached Erhard Milch suggesting that it might be a good idea to let combat pilots fly the Me 262 and evaluate it without any pressure from manufacturers or other quarters. Adolf Galland was a 'down-to-earth' pilot and not interested in the political in-fighting that engaged the majority of senior Luftwaffe officers. He was also a hands-on leader and not one to let his men fly anything that he thought dangerous simply for the sake of expediency. Erhard Milch knew this and sent Galland to Lechfeld to fly the aircraft and report back to him. Arriving at Lechfeld, Galland was briefed by Fritz Wendel and the other test pilots. Willy Messerschmitt showed him around the aircraft and explained about the engines and gave him a cockpit check. Galland then eagerly clambered into the cockpit of the V3. After a second cockpit check the ground crew strapped him in and started the first engine. As it crackled into life, Galland felt the power start to surge through him and the aircraft. The second engine burst into life – and then burst into flames. The ground crew hurriedly got Galland out as they extinguished the fire. Fortunately the V4 had recently joined the programme, so that was wheeled out instead for his test flight.

Within minutes of clambering into the cockpit he was racing down the runway and was airborne. The rate of climb left him breathless as he sped upwards, then as he got used to the feel of the aircraft he spotted another aircraft on a test flight, the large Me 264 long-range bomber. Realising that this would be the ideal opportunity to use the aircraft for its intended purpose as an interceptor, Galland carried out a high-speed interception pass of the Me 264. Galland was sold. Work on the aircraft in his opinion needed to be advanced quickly so that it could be in operational service as early as possible. He wrote to Erhard Milch extolling the virtues of the aircraft by saying that in his considered opinion this was the aircraft that would take them forward and not on the defensive.

On 25 March 1943 a plan for the production of the Me 262 fighters was placed before the RLM, with strong recommendations for bomber and reconnaissance versions. Within the plan there was a recommendation for the production of the Me 209 fighter, a derivative of the Bf 109 but built purely for speed and of no real combat value. Galland saw that this was just a propaganda ploy and that the aircraft had no future as a fighter and argued, successfully, for the project to be abandoned. Three days after the RLM had received the production plan, they ordered that 100 production aircraft be built and at a meeting on 2 June 1943 of the *Generalluftzeugmeister-Besprechung* (Chief of Procurement and Supply Conference) orders to release the design and materials were issued. In theory this was ideal, but in reality the situation regarding supplies and materials was worsening and it was to be April 1944 before full production of the Me 262 started. In the meantime development work continued on the aircraft because despite all the accolades that aircraft was receiving, in reality it was not ready for operational service. There were continuing concerns about the reliability of the engines and the development of the ejection seat (using the Me 262 V3 as a test bed), was continually being postponed. The first flight of the fifth prototype, the V5 (PC+UE) took

Me 262s of Kommando Nowotny on the flight line.

A captured Me 262B trainer about to be evaluated by American test pilots.

place on 6 June 1943 and it was fitted with a tricycle undercarriage. The prototype was fitted with the nosewheel in a fixed position but the initial tests were disappointing. It had been hoped that the take-off run would have been shortened but it was the same length as with the original undercarriage. It was decided to fit two solid propellant Borsig RI-502 booster rockets either side of the centreline beneath the fuselage just behind the centre of gravity. Test pilot Karl Baur flying the aircraft had to keep the nose down almost immediately the rockets and engines were ignited. To counteract this, the rockets were moved forward and angled downwards and this reduced the take-off run by 800 feet, a substantial difference.

Modifications to the Me 262 V1 gave it a pressurised cockpit and the latest Jumo 004A turbojet engines. It was also fitted with three MG 151 cannons mounted in the nose and carried out its first flight on 19 June 1943. The summer months brought an increased supply of the Jumo 004B-0 engines delivered to the Messerschmitt facility for flight-testing. Increased interest in the Me 262 brought the Luftwaffe's hierarchy to Rechlin for a demonstration. Among these was Hermann Göring who, having had a verbal report from Adolf Galland, wanted to see for himself this remarkable aircraft. The demonstration flight took place on 25 July 1943 flown by test pilot Gerd Lindner in the Me 262 V4. Göring was impressed and immediately saw it as a multi-role aircraft that he knew would please Hitler.

The following day, whilst Göring was on his way back to Berlin, Gerd Lindner was flying the aircraft back to Lechfeld when he diverted to the airfield at Schkeuditz. As he approached there was an engine failure and the aircraft crashed on attempting to land. The aircraft was written off but Gerd Lindner was unhurt. To make things worse, just over a week later, on 4 August 1943, the V5 on its next test had the nosewheel collapse whilst the aircraft was taxiing, causing extensive damage to the nose section. It was not until January 1944 that the aircraft was back on line and one month later it was almost written off when on landing; one of the tyres on the main undercarriage blew causing extensive damage to the aircraft. The aircraft was not repaired before the end of the war.

This left two Me 262 prototypes, the V1 and V3, to carry out all the tests until the V6 was ready to join them. That was going to take another two months.

Projekt 1065

At the end of September 1943 the RLM produced a document entitled *Projekt 1065*. In the document they laid out the programme they expected the company to follow with regard to the Me 262. There were to be two versions of the photo-reconnaissance model – one to be fitted with one RB 75/30 camera, the other with two RB 75/30 cameras; two high-speed model bombers, the first to carry one 2,220lb or two 1,100lb bombs and the second model was to carry a similar load but would contain different equipment. There were also to be three interceptor fighter versions. The first, to be armed with six 30mm MK 108 Rheinmetall-Borsig cannons and fitted with one Walter HWK RII/211 supplementary rocket motor; the second to be armed the same but fitted with two Walter HWK RII/211 supplementary rocket engines; and the third was to be a two-seat trainer fitted with four MK 108 cannons. All the aircraft with the exception of the second version interceptor were to be powered by the Jumo 004C engines, the second interceptor was to be powered by two BMW 003R engines.

'Wonder Weapons'

With the war not progressing as they had hoped, the Nazi hierarchy started to take an even greater interest in the development of the jet fighter and bomber. They saw these aircraft as the one thing that could take the fight to the Allies. In Berlin, Joseph Göbbels was incensed that in his opinion there were more workers in Messerschmitt's design bureau than there were on the shop floor. Messerschmitt, in his opinion, was more interested in furthering the science of aviation than he was in winning the war. This of course was utter nonsense as Willy Messerschmitt was committed to the Nazi cause and had spoken out in support on a number of occasions, including a radio broadcast. This came at a time when the raids on Germany were becoming more and more frequent and with increasing numbers of heavy bombers. Göring and other senior members of the Luftwaffe had been summoned to a meeting with Adolf Hitler and had been subjected to a tirade of criticism and abuse that lasted more than five hours.

Göring managed to persuade Hitler that it was not the fault of the Luftwaffe but of the manufacturers, who were not meeting the demands placed upon them. Hitler, still not sure about Göring, summoned the major manufacturers to a meeting in his mountain retreat on 27 June 1943. Those invited were Dornier, Heinkel, Arado, Henschel, Blohm & Voss, Focke-Wulf, Junkers and Willy Messerschmitt. It is interesting to note that Hitler had invited Willy Messerschmitt to the meeting even though Messerschmitt had been deposed as chairman of the company some time earlier. Each of the manufacturers was invited to an individual private meeting with Hitler and given the unusual opportunity, at his insistence, to be open and frank about their concerns about the way the manufacturing process was being constrained by the bureaucracy of the Luftwaffe and the RLM.

The meeting with Willy Messerschmitt lasted over an hour, during which the development of the Me 262 and the abandonment of the Me 209 was discussed at length. Hitler decided to continue with the Me 262's development, but to keep the Me 209 on the back burner, so to speak.

Camouflaged Me 262 bomber being prepared for a bombing mission.

Schweinfurt–Regensburg missions

Then came one of the severest blows to the German aircraft-manufacturing programme. For some time the Allies had known that the giant Messerschmitt production and assembly plants at Wiener Neustadt and Regensburg produced almost half of the fighter aircraft of the Luftwaffe. Because the plants were situated in the eastern part of Germany they were protected from raids because neither the Americans nor the British had bombers with that range. With the defeat of the Afrika Korps in North Africa, the Allies now had bases there. It was decided to attack the plants in two ways. Bombers from England, after dropping their bomb loads, would continue on to North Africa, whilst bombers from North Africa would drop their bomb loads and return to their bases. The raids had to be a complete surprise and because of the distance, the bombers would have no fighter cover over the targets. Initially it had been decided to attack both plants simultaneously, but because of poor weather the decision was made for two separate raids. These raids were to be part of the largest bombing mission in the history of the USAAF.

The attack on Wiener Neustadt took place on 13 August 1943 when 65 Boeing B-17 Flying Fortresses and B-24 Liberators from the 9th Air Force took off from airfields near Benghazi and flew 1,200 miles to their target. They encountered very light opposition from both aircraft and anti-aircraft fire and extensively damaged the production lines at the plant. The attack on Regensburg was to be combined with an attack on the ball-bearing works at Schweinfurt. The 230 B-17 Flying Fortresses, split into two task forces – one of 116 the other of 114 – on the Schweinfurt-Regensburg mission were fitted with special long-range fuel tanks. The attack was devastating to the Germans, both plants were damaged and production almost halted. The cost to the Allies for the mission was 147 B-17s lost or damaged and 550 crewmen killed or captured. The ball-bearing factories at Schweinfurt had ground to a halt after the bombing and the production line at Regensburg had been halted, albeit only temporarily. Four hundred aircraft had been destroyed as well as the equipment. The Me 262 production line and jigs were destroyed totally, setting the programme back even further.

'Revenge' of the *Schnellbomber*

The raids stunned Hitler and drew an angry response demanding a rapid retaliation. The bomber force of the Luftwaffe was brought to a rapid state of readiness. Willy Messerschmitt saw the opportunity to take advantage and approached Hitler personally with regard to the Me 262 as the *Schnellbomber* (Fast bomber). He pointed out to the Führer that Göring himself had supported the Me 262 and that all the fighter pilots who had flown the aircraft also supported it. Messerschmitt was well aware of the fragile relationship that Göring and Erhard Milch had with Hitler and the incompetence that had led to it, and wanted to capitalise. He also had a long-standing score to settle with Milch, so if he could get Hitler on his side there might be a chance to get even.

Pilot about to climb into the cockpit of his Me 262.

A hurried conference at Berchtesgaden was called on 14 October 1943 (hosted by Göring) to which Willy Messerschmitt was invited. Its purpose was to decide on the bomber that was to be the instrument of the reprisals that Hitler had demanded from Göring. There were three aircraft in the frame: the Me 264, Me 209 and the Me 262. Time was running out and the development of the Me 264 was going to take too long even though there were models available, so that was discounted. This left the Me 209 and the Me 262. The Me 209 up to this point had not proved satisfactory, which left the Me 262. Hitler, in the meantime, having been briefed thoroughly on the Me 262 by Willy Messerschmitt, put forward his thoughts on the aircraft as if they were his own. Göring and Milch argued on the side of caution, but eager to curry favour with the Führer, were not too forceful. Göring, having once denounced his fighter pilots for cowardice becuae of the number of effective bombing raids that were getting through, now needed to get them on his side. So he approached his *General der Jagdflieger* Adolf Galland for support in pushing for an increase in fighter production, to include the Me 262. Allied bombing raids had taken their toll on the fighter production lines and there was very little left in the way of jigs and tooling to carry out a massive production programme, especially for the Me 262. In an effort to appease Hitler, Göring suggested that he was aware of the need for a fast bomber and that the way forward would be to utilise the Me 262 programme by producing a bomber version of the aircraft.

At a meeting on 2 November 1943 Göring visited the Augsburg facility for a meeting with Messerschmitt, Galland, Milch and Vorwald, who was head of the *Technische Amt.*

Göring, unaware that Messerschmitt had had the ear of Adolf Hitler for the past two months regarding the development of the Me 262 as a bomber, reiterated the Führer's thoughts on the production of a fast bomber. As if hearing the suggestion for the first time, Messerschmitt said that it would only take a couple of weeks to convert the Me 262 fighter into a fighter/bomber. This, in effect, would satisfy not only the fighter pilots, but also those who wanted a fast bomber. First, however, the ongoing engine problem that had dogged the project from day one had to be resolved and Messerschmitt approached Dr. Anslem Franz, Director of the Jumo programme, who admitted that the Jumo 004 engine was in no way ready for mass production.

In an effort to get final approval for the programme, a display of the only two Me 262s available, V4 and V6, was given at Insterburg to Hitler and Göring on 26 November 1943. Flown by test pilots Gerd Lindner and Karl Bauer, they put on a display that satisfied Hitler that Messerschmitt had been telling him the truth at their meeting on 7 September.

There was no signing of documents, there were no words passed between Adolf Hitler and Willy Messerschmitt, but it was accepted that the Me 262 programme was up and running and had the sanction of the Führer. The time for recriminations was over, it was time to get the production lines rolling. Pressure was put on Junkers to get the Jumo 004 engine into production. In the meantime the Messerschmitt factory started to get the jigs and tools ready. The next prototype, the V7, appeared from the factory on 20 December 1943 and was immediately flight-tested by test pilot Gerd Lindner. The aircraft was powered by two Junkers Jumo 004B-1 turbojet engines and had a clear plastic canopy that faired into the fuselage in place of the framed one and had a tricycle undercarriage. It was also the first to be fitted with a pressurised cockpit that was tested up to around 39,400 feet.

With the factory lines now starting to get under way it was decided to create a test detachment/operational-training unit that would be able to test the aircraft as they came off

Captured Me 262 being towed by an American Jeep to be evaluated.

the production line. On 9 December 1943, *Hauptmann* Werner Thierfelder, a South African with German parents, had come from *Zerstörergeschwader ZG 26* 'Horst Wessel' and was given the task of setting up the formation of the *Erprobungskommando 262* (Ekdo 262). He had no aircraft, no pilots and no ground staff. At first he used the company test pilots as instructors until he himself had learnt to fly the Me 262. It was not until 21 December that he took the V6 prototype on a familiarisation flight.

By the beginning of January 1944 the first of the Ekdo 262 test pilots had been trained up but still there were no aircraft to test.

The Me 262 Prepares for War

As if the Allies had been made aware of this progress, on 25 February 1944 they carried out another series of raids on the Messerschmitt plants at Regensburg, Augsburg and the ball-bearing works at Schweinfurt. In order to achieve the maximum damage the bombers were ordered to fly at a greatly reduced bombing altitude. The raid caused extensive damage to the plants and dramatically reduced production of the fighters being built there and it was to be many months before they were able to recover. Unbelievably, some production of aircraft continued despite the raids; but in the air the Luftwaffe was also losing aircraft and, more importantly they were losing pilots. In does not matter how many aircraft exist, if there are no pilots to fly them then they are just useless pieces of metal.

The whole Me 262 programme was moved away from Augsburg to Leipheim airfield where the final assembly was carried out. Because of the ever-growing intensity of the Allied bomber raids on the aircraft manufacturing plants it was decided to create large concrete bunkers away from the airfields and assemble the aircraft there. It was also decided to break the companies down to over 700 small units and disperse them all over southeast Germany. Known as *Kunos*, a number of these facilities were built in the forest around the airfield at Leipheim and close to the Munich-Stuggart autobahn. The latter had a central straight stretch of road of 6,560 feet, which could be used as a runway. When aircraft were assembled they could be flown out to other airfields for flight tests before delivery.

Captured Me 262-1a with a 50mm cannon in American livery.

The Allies became aware of this and carried out numerous photo-reconnaissance missions in an effort to locate them. In one raid on Kuno I (a facility that assembled the aircraft ready to fly) on 24 April 1944, Allied bombers destroyed over fifty Me 262s that were waiting to be delivered to the Luftwaffe. By the end of the year two other facilities, Kuno II and III, were fully operational and they, like Kuno I, were situated close to the autobahn. By the end of March 1944 almost all the major factories had been dispersed into small satellite companies that lay scattered across Germany. The only weak point as far as the Luftwaffe was concerned was that there had to be some large facilities where the components could all be brought together and assembled into aircraft. Once that had been done there had to be an area where the aircraft could be parked and facilities to test fly them before they were delivered to the Luftwaffe. This meant open space. It was these areas that the Allied photo-reconnaissance aircraft were looking for when they carried out high-altitude flights over Germany.

One idea put forward was to build giant assembly plants which would be mostly underground. The idea was approved and in March 1944 construction began on a building 1,310 feet long, 151 feet high, (half of which would be below ground), and 279 feet wide. It was to have an arched concrete roof that went from being 10 feet thick at its lowest point to 59 feet thick at its highest and could withstand multiple impacts of up 22,000lb. When completed it was to be covered with earth and made to look from the air like a gentle rolling hill. The building consisted of five floors, the two lowest for assembling the smaller parts of the aircraft, the next two floors for housing the several hundred workers that assembled the aircraft and the top floor, where the final construction of the aircraft took place, would also double as a runway. To use the short runway the aircraft were launched with the aid of rocket-assisted motors.

Over 30,000 people were put to work constructing the *Weingut II*, as it was codenamed. The workforce was all slave labour and half came from the Dachau concentration camp, which was situated 37 miles from the site. New campsites were set up five miles from the site and the inmates were made to walk to the site, work a fourteen-hour shift and walk back. The conditions they had to work under were appalling; little food or clothing and little or no medical facilities. Before the end of the war over 10,000 Jews and other prisoners had died in constructing the Weingut II. The project was never finished and although the 'hangar' was put to some use it never fulfilled its potential.

The Allied raids were having their effect on production of the Me 262 but by the end of February 1944 the first tests using the V8 prototype were carried out using four 30mm MK 108 cannons. The loss of another of the prototypes, the V6, on 9 March 1944, was the cause of some concern. An engine problem resulted in the aircraft crashing and killing the pilot Kurt Schmidt. This left just four prototypes still flying, V1, V3, V & and V9, but they were soon joined by the V8. On 18 March it carried out the first air-to-air firing tests of the nose-mounted cannons. The jubilation was short-lived, as the following day during a routine test flight the V7 appeared to stall then plunged into the ground, killing the test pilot *Unteroffizier* Hans Flachs. This of course reduced the prototypes to four once again. The last of the ten scheduled prototypes, the V10 (VI+AE) appeared from the Augsburg-Haunstetten facility on 15 April 1944. This had been built with all the other prototype weaknesses taken

Bomber version of the Me 262.

into consideration – it was the Me 262 that would be flown to the limits of its envelope. It was also the first one to have been converted for the fighter/bomber role.

The long awaited fighter/bomber that had been so eagerly waited by senior Luftwaffe officers was the Me 262A-2a *Sturmvogel* (Assault Bird). Immediately the first *Kampfgeschwader*, KG 51, was equipped with the fighter/bomber and formed a *Geschwaderstab* (Staff Group) based at Rheine commanded by *Oberstleutnant* Wolf-Dietrich Meister. The group began sending pilots to *Erprobungskommando* Schenk at Lager Lechfeld under the command of Major Wolfgang Schenk. The conversion was not easy for pilots who had learned to fly on piston-engined aircraft and the difficulties they experienced showed in the length of time it took to get the first *Gruppe* of KG 51 up and running.

Even then trials were still being carried out using the Me 262 V10 trying to establish the best range of bombs that the aircraft was able to carry. The *Deichselschlepp* towed bomb was one under consideration. This was a single 1,000lb bomb fitted with wheels and towed behind the Me 262 by means of a fixed steel bar. The project was abandoned after tests because of the instability of the towed bomb and handling difficulties experienced by the pilots when towing it.

A desperate urgency characterised the efforts of the Luftwaffe's senior officers to get the Me 262 into operational service. Already there were bitter arguments springing up between those who favoured the Me 262 as a fighter, such as Galland, and those who saw its role as that of a fighter/bomber. Hitler, in the meantime, was demanding that the aircraft be a blitz-bomber and instructed Göring to resolve the argument between his fighter pilots and his bomber pilots. Göring, desperate to get back into the the Führer's good books, simply told a gathering of his most senior Luftwaffe officers that Hitler had issued a direct order that the Me 262 was to be a fighter/bomber. Galland was furious, he believed that the future of the Me 262 lay as a pure fighter and it should not be limited to carrying a single bomb. In order to placate his *General der Jagdflieger* Galland, Göring promised that the fighter programme would run alongside that of the fighter/bomber, although it was obvious that this was going to be extremely difficult. Hitler, surprisingly enough, announced this decision the following day, probably owing more to the efforts of Willy Messerschmitt than Hermann Göring – the latter, not so surprisingly, claimed to have persuaded the Führer to follow this line of thinking.

Despite all the arguments and so-called problems, the main reason for the delay in making the Me 262 operational was the production of the Junkers Jumo 004B turbojet engine and its configuration for operational use. The problem was in fact never to be resolved satisfactorily.

Four pre-production models were delivered from the factory, S1, S3, S5 and S6. The following month a further eleven production models, followed by 28 in June and 58 in July were delivered. On 10 June 1944 the first Me 262 arrived at *E-Stelle* (codename for the Rechlin Test Centre) and preparations for training operational pilots got under way. The training was intensive and by September they had accumulated over 350 hours in 800 flights. This would have been more but there were problems with the undercarriage. The first operational unit *Erprobungskommando 262* had received its first operational Me 262, the V8, back in April 1944 and a number of pilots from III/ZG 26 carried out their initial training with them. The pilots had their first experience in a converted Bf 110 and a Me 410. Pilots who had flown twin-engined aircraft found the conversion far easier than the single-engined fighter pilots. High-altitude flights and flying in *Rotten* (pairs) culminated in gunnery practice and simulated dogfights. The first course consisted of 14 pilots, ten remained with the unit and two of them were given command of the two *Staffeln* within the unit.

The first bomber unit to receive the Me 262A-2a was the III/KG(J)(*Kampfgeschwader* (Jet) 51 who went to Lechfeld to collect the aircraft. It was there that they went through the conversion training and bombing practice. Because of the slow rate of production, the number of aircraft available for instruction was relatively small, limiting the number of hours and flights for each pupil. This was further aggravated when the aircraft were constantly being diverted to testing duties at other airfields or facilities. Despite this a large number of pilots were trained up to fly both the fighter and the fighter/bomber and the first recorded encounter in combat came on 26 July 1944. A photo-reconnaissance Mosquito of No.544 squadron RAF, flown by Flight Lieutenant A.E.Wall and Pilot Officer A.S.Lobban, were on a mission over Münich when they sighted an aircraft closing on them from the rear. Opening the throttles of the Mosquito, Flt.Lt.Wall felt the surge of power race through his aircraft as he raced away, confident that the Germans had nothing that could catch him. Within seconds the German aircraft had closed and then screamed past. Wall realised that he was now being attacked by one of the new German jet fighters that he had heard about and took evasive action. All to no avail as the next minute the jet was back on his tail firing at him. Breaking to starboard he lost the jet temporarily but realising that there was no way he could outrun the jet, he headed for cloud cover. When he emerged the jet was gone, but during the violent manoeuvres of his aircraft a side panel had come off and damaged his tail. Wall headed for the nearest friendly airfield, which was in Italy, and landed safely. The jet had been flown by *Leutnant*

Alfred Schreiber who returned to his base desperately short of fuel and claiming the destruction of one Mosquito.

At the same time III/KG(J) 51 known as *Kommando* Schenk after its commanding officer at Châteaudun, France, was preparing its Me 262A-2a for its first mission on the Allied beachheads. On the face of it this was to be a new chapter in bomber warfare but it was only a mild incursion and it achieved almost nothing. The unit reported back to *I Gruppe* at Rhiene on 5 September. Efforts to get the Me 262 up to a state of combat readiness continued with Ekdo 262 and in an effort to appease the Luftwaffe High Command, a number of claims were made with regard to successes against Allied aircraft. None of these claims was ever substantiated, but one significant success was enjoyed on 15 August 1944. A lone B-17 Flying Fortress returning from a raid over Stuttgart was intercepted by a *Rotte* of Me 262s from Ekdo 262. The two jet fighters, flown by Feldwebel Lennartz and Ofw Kreutzberg closed on the B-17 so fast that the gunners on the bomber had no chance to open fire. With the first burst from his 30mm cannon, Lennartz sliced off the bomber's port wing sending it tumbling out of control into the ground. By the end of August a number of other claims of success were made, including a Spitfire and a P-38 Lightning. But the Luftwaffe was not having it all its own way. In the last week of August 1944, nine Me 262As were sent from Lechfeld to reinforce III/KG(J)51. Two crashed on take-off, one developed engine problems and crashed and one had to force-land in a field due to engine failure. The remaining five arrived safely but they were plagued with minor problems.

Two days later *Obfw.* Lauer on a patrol over the Dendermonde area of Belgium came across some ground-attack aircraft carrying out low-level attacks on German troop movements. Thinking he had easy targets he raced into attack, unfortunately for him he had not looked around and failed to spot eight P-47 Thunderbolts of the 78th Fighter Group, US 8th Air Force from Duxford, who were flying cover for the low-level attack bombers. They however had not failed to spot him and Major Joe Myers and his wingman Lieutenant Manfred Coy went after him. *Obfw.* Lauer suddenly realised that he was under attack and in an extremely vulnerable position and carried out a series of low-level manoeuvres in an effort to shake off his pursuers. As he did so his starboard wing touched the ground and the aircraft cartwheeled into a field. Lauer jumped from the aircraft unhurt, and raced for cover just before the P-47s machine-gunned his aircraft destroying it completely. This incident highlighted the lack of combat experience that a number of the surviving Luftwaffe pilots had. All those who did have the experience were desperately fighting on other fronts, leaving the gaps elsewhere to be filled with comparative novices.

In Berlin, the Chief of Staff of the Luftwaffe, *Generalleutnant* Werner Kreipe was pressing for the Me 262 to be assigned to a day fighter role. This infuriated Hitler as he had from the start insisted that the Me 262's role was to be that of a *Sturmvogel*. After bitter arguments he relented far enough to agree that 5 per cent of Me 262 production could be Me 262A-1as, whilst the remainder were to be Me 262A-2a *Sturmvogel* fighter/bombers. Kreipe was not satisfied and continued to push for the Me 262 to be a pure fighter aircraft. The arguments became acrimonious, and on 19 September, after one extremely bitter argument, Kreipe resigned. General Karl Koller, who by now was no more than a figurehead, took his place.

Remains of a concrete hangar used to house the Me 262.

The Luftwaffe was now excluded from the military conferences in Berlin with Hitler. Göring was only summoned to Berlin to be berated by an increasingly unstable Hitler. The military infrastructure was coming apart at the seams.

Ekdo 262 continued to operate and by the end of September 1944 claimed to have shot down and destroyed three Mosquitos, one Spitfire and a P-51 Mustang. The use of the Me 262 was further boosted when at the end of September 1944, Adolf Galland ordered one of the Luftwaffe's top aces, Major Walter Nowotny, who had 255 kills to his name, to form a test unit. The unit, known as *Kommando Nowotny*, drew pilots from Ekdo 262 and from III/ZG26 and they were based at Achmer. By the beginning of October the unit was operational with some 25 Me 262s. On 7 October they carried out their first operational assignment when they were scrambled to intercept massed formations of B-17 Flying Fortresses who were attacking oil installations. There had been earlier concerns about the vulnerability of the Me 262 when taking off and before it reached it operational speed, and plans had been agreed to put a number of Focke-Wulf Fw 190s in the air to give the Me 262s cover. On this occasion their fears were realised when five of their aircraft were 'bounced' as they were taking-off by P-51 Mustangs from the 361st Fighter Group who were escorting the bombers. Two of the Me 262s were shot down as they were lifting off the runway, a third was caught as it tried to climb away and a fourth whilst it was attacking the bombers.

Protection for the Me 262 was stepped up and Focke-Wulf Fw 190s from 9 and 10 *Staffeln JG54* were assigned to *Kommando Nowotny* to fly for cover for the Me 262s during the critical periods of take-off and landing. On 15 October 1944, the first test of this proposal occurred, when six Fw 190s from *9 Staffeln JG54* took off from Hesepe to fly cover. The idea was that the Me 262s would only take about six minutes to take off and get into operational mode, but as the jets started to climb away, the Fw 190 pilots suddenly realised that there was a force of some 40 P-51 Mustangs from the 78th Fighter Group, US 8th Air Force, heading their way. Minutes later they were attacked by the P-51 Mustangs and five of the six Fw 190s were shot down. The sixth managed to escape but was later shot down during another mission.#

Better Armament

In an effort to improve the odds, *Kommando Nowotny* carried out trials using air-to-air rockets. The two forward mounting points for bombs were each fitted with a single tube that could carry a single 210mm rocket. These rockets were fitted to six Me 262s and an attack was carried out on a formation of B-24 Liberators who were carrying out another attack on the oil installations escorted by P-47 Thunderbolts. The Me 262s fired their rockets all at the same time, but there is no record of one hitting its target. It did however shake the bomber crews and their fighter escorts, prompting the 392nd Bomber Group to carry out their own unauthorised experiment with rear-firing bazookas.

Trials were also carried out using a massive 50mm Mk.108 cannon mounted in the nose of the Me 262. Two aircraft were modified to take the cannon, which was easily identifiable by the huge barrel sticking from the nose and the hump immediately behind. The trials revealed serious deficiencies in its use and, coupled with the complaints from the pilots who flew that aircraft during the tests, that was enough to cancel the project.

Captured Me 262s aboard a British aircraft carrier being taken to the UK.

The First Jet Fighter Unit

On 7 November *Kommando Nowotny* suffered a serious setback when the commanding officer Major Walter Nowotny, was shot down and killed. Nowotny had intercepted a radio message from a crippled B-24 Liberator stating that he had engine trouble and was returning to base. On his way to find the B-24 Nowotny was jumped by a P-51 Mustang and raked with gunfire. Adolf Galland, who was visiting the unit at the time, watched in horror as suddenly he saw Nowotny's Me 262 break through the low cloud in a vertical dive and then smash into the ground. Nowotny's replacement, *Hauptmann* Georg-Peter Eder, was appointed almost immediately, before the unit even had time to mourn Nowotny's passing. It was a short-lived appointment as three days later the unit was absorbed into II/EJG2 and moved back to join the unit at Lechfeld. Just before the unit was absorbed into II/EJG2, *Hauptmann* Eder shot down three B-17 bombers making him one of the top jet fighter aces of the Luftwaffe.

The formation of *JG7 Geschwaderstab* at the beginning of August 1944 created the first fully equipped operational jet fighter unit. The unit had been designed primarily for the defence of the Reich, to which the Luftwaffe was now committed. Such was the scale of the mass bomber raids against Germany that any thoughts of taking the war to the Allies had long gone. Raids on England were few and far between and the bombers were now being used to fly above the Allied bomber formations and drop bombs on them in the hope of hitting some of them. The Luftwaffe was now in desperate straits: their losses in 1944 were the highest of the war – 27,000 plus. The aircraft factories were working at full pace to cope and along with the demand for new aircraft came the demand for trained pilots. Fuel shortages were also causing serious problems, especially for the jet aircraft.

The desperate need for information on the Allied advances highlighted the lack of a special reconnaissance unit and so the *Sonderkommando Braunegg* was created at the end of November 1944. The new unit under the command of *Oberstleutnant* Braunegg was equipped with Me 262A-1a/U3s and a number of modified Me 262A-1as. Although the unit was operationally ready by the beginning of December, because of the appalling weather it was to be the end of December before the first mission was undertaken.

Last Throw of the Dice in the Ardennes

Losses of the Me 262 were still unacceptably high. In November 1944 21 jet aircraft were lost either through mishaps or engine malfunctions and a small amount were lost in combat. There were continuing problems with engines, nose-wheels and tyres. Because of the more or less helpless situation in which the Luftwaffe found itself, very little time could be spent in carrying out tests to find the causes.

At the end of December one of the last major offensives of the war took place – the Battle of the Bulge, or the Ardennes offensive. At dawn on 16 December 1944, German artillery positions opened fire along the Front and Panzer divisions made a mass armoured assault on the Allied positions. The Allies were caught unawares and along the Front German tanks broke through the lines and headed toward Antwerp. The Luftwaffe had assembled 2,460 aircraft, of which almost 2000 were fighters, under the command of Adolf Galland to support the counter-offensive known as *Bodenplatte* (Baseplate). The weather, however, had closed in and the aircraft could not get off the ground to support the attack. The Allied aircraft were also affected by the weather and could not attack the German ground troops as they pushed their way through.

The weather lifted enough for Allied photo-reconnaissance aircraft to get into the air and Me 262 fighters were launched to intercept them, which they did with some success. For some unknown reason the bulk of the Luftwaffe's aircraft was held back and when the offensive fizzled out, the overstretched supply lines came under constant attack from Allied fighters. On the ground American troops halted the advance, consolidated their positions and then began once more to push the beleaguered German troops back. The Luftwaffe was ordered onto the offensive and the Arado 234 jet bomber made its first operational appearance with repeated sorties against the American army. The Me 262A-2as of III/KG(J)51 made repeated ground attacks on American tanks as they pushed their way forward, and lost only one aircraft. The respite in the weather also brought the Allied fighters and bombers out in force and in their overwhelmingly superior numbers they attacked the German airfields. By the end of the first day of the American counter-offensive, the Luftwaffe had lost 85 pilots and over 100 aircraft. As the Americans pushed on farther and farther, the Luftwaffe was reduced to nuisance raids and ground attacks. The Me 262 fighters were now carrying out their strafing attacks at maximum speed of 560 mph, which made it extremely difficult for the Allied fighters to intercept them.

The Allied fighter pilots, however, were learning to get the measure of these jet fighters. One disadvantage the jets had was that when attacking at high speed all turns had to be made in long sweeping moves, and any tight turns reduced the aircraft's airspeed dramatically. As a high-speed interceptor or a ground attack fighter the Me 262 was unequalled, but in a dogfight it was no better and probably worse than the piston-engined fighters.

All over the Christmas period and right up to the start of the New Year, the fighting was intense and increased production brought more and more jet fighters to the units. The problems with the aircraft and the training of new pilots continued, with the result that a large proportion of the losses suffered by the units was not due to the enemy but to pilot error and malfunctions of the aircraft. This was not helped by the infighting that continually plagued the Luftwaffe hierarchy. In January 1945 an attempt to remove *Feldmarschall* Göring as head of the Luftwaffe was made by senior members of the Air Staff. Dissatisfaction raced through the Luftwaffe, as no one appeared to know who was in charge and who was making the decisions. At the end of December Gordon Gollob, a personal friend of Heinrich Himmler – and a sycophant towards Hermann Göring – relieved Adolf Galland of his post.

With Galland's dismissal, doubts surfaced in the SS with regards to his loyalty to the Führer and rumours abounded that he could be tried for treason. But Galland still had some very powerful friends and one of them rang Albert Speer, who in turn spoke to Hitler. Speer was a close confidant of the Führer and the SS were warned by Hitler personally to withdraw any such accusations they may have harboured against Galland.

Göring was furious and ordered Galland to meet him at his country estate, where for an hour he attacked Galland over his tactics and integrity. Such was the state of the Luftwaffe that Galland immediately put forward a proposal to set up his own *Staffel*-sized fighter unit, a move that had earlier been dismissed by Göring. Göring, realising that Galland obviously had powerful friends, conceded and agreed that he would be independent from Gollob's control.

Galland Independent

Galland's new unit was called *Jagdverband 44* (JV 44); but it was a unit without men and aircraft. Word spread like wildfire that the legendary former *General der Jagdflieger* was looking for pilots and ground crews and soon Galland was inundated with requests to join him. The aircraft the unit wanted was the Me 262, and Galland acquired the aircraft from III/EJG2 and 16/JG54. By the end of February 1945 JV 44's request for 16 aircraft and 15 pilots for the unit, and 7 aircraft and 7 pilots for the *Stab* had been answered. One of the first pilots Galland requested was Major Johannes Steinhoff who was fast becoming a legend amongst the fighter pilots. Between them they set up the unit and started to recruit the right personnel.

Other units that were equipped with the Me 262 were still fighting a tenacious rear-guard action. On one raid on 3 March 1945, over 1,000 heavy bombers from the 493rd, 467th and 445th Bomber Groups, escorted by more than 700 fighters, were intercepted by Bf 109s and Me 262s near Hanover. Twenty-six Me *262s* went into the attack racing through the formations firing burst after burst. The turret gunners in the B-17s opened fire as they saw the jets diving from behind and occasionally managed to hit one of them. The escorting P-51D Mustang fighters were having extreme difficulty in containing the jets. One American pilot, Lieutenant Marvin Bigelow from the 78th Fighter Group, chased a Me 262 in a dive with his throttle wide open, attaining a speed of almost 600 mph. According to his report the wings were vibrating so much he thought they were going to come off. The Me 262 easily outstripped him and the P-51D was pulled out of the dive and returned to escort duties. On his return to England the aircraft was examined and found to have been stressed so badly that it was beyond repair.

The attack on the Allied raid was not the success hoped for, seven bombers and two escort fighters were shot down, but six Me 262s were lost together with a number of Bf 109s. The Luftwaffe bomber units were having a horrendous time, losing large numbers of their aircraft to Allied fighters. On 21 March 1945, I/KG(J)51 lost four jet bombers and I/KG(J)54 lost two. The new I/KG(J)51 *Gruppenkommandeur* was one of the pilots who was lost when his bomber was shot down.

Galland's JV 44, based at the Brandenburg-Briest airfield, was still in the process of 'working up' as more and more pilots were being trained in flying the Me 262 jet. The western Allies and the Russians had by now pushed back the German Army to such an extent that the only action it and the Luftwaffe could muster was rearguard. Galland's JV 44 was ordered into action, and whilst Steinhoff trained the pilots Galland harassed the authorities for more aircraft. He also developed a number of new techniques for his pilots, instead of flying in *Rotten* (two-aircraft sorties) or as a *Schwärme* (Flight), they would fly in *Ketten*. This meant that three aircraft would fly in a triangular formation approximately 450 ft apart.

Just when Galland thought things were starting to improve Hitler appointed *General der Waffen-SS Obergruppenführer* Hans Kammler as head of Jet Aircraft Production and Operational Deployment. This meant that Kammler had control of everything that had to do with jet aircraft. Just one more irrational decision made by Hitler.

Rumblings amongst the military hierarchy after the failed assassination of Adolf Hitler at Rastenburg on 20 July 1944 hinted at another attempted coup against the Führer. The military leaders could see the way the war was going and realised that if the Russians were to invade Germany they would be under communist occupation. Thoughts turned to approaching the British and American commanders with a view to striking a deal with them and then to join forces against the communists. Nothing was ever done, mainly because of the fanaticism of the Nazis that still held positions of power and authority.

The end of March saw the whole of JV 44, with the exception of the aircraft, packed aboard a train heading south toward Munich-Riem. By this time Galland had just 12 operational aircraft and less than 20 operational pilots, including himself and Johannes Steinhoff. Attempts were made to disband the unit, but things were now so desperate that no one even pursued the suggestion. On 7 April a mass raid of 2,000 heavy bombers pounded the industrial targets, the railway yards and retreating troops. Germany was on the receiving end of over 5,000 tonnes of bombs per day and the aerial defences, although tenacious, were having little or no effect against the overwhelming odds. Equal numbers of fighters were escorting some of the heavy bomber raids.

The losses of Me 262s continued to mount, and on 10 April 1945, JG 7 were called into action when 442 bombers from the 1st Air Division were spotted heading for Rechlin. Fifty-five Me 262A-2as were put into the air to intercept. As they approached the formations they noticed that they were unescorted and tore into the attack in *Rotten*. Within minutes five B-17s were heading for the ground on fire, but then, as they made their second pass, fighters from the 20th Fighter Group arrived, resulting in six of the Me 262s being shot down and a number badly damaged. Seven days later P-47 Thunderbolts carried out a strafing and ground attack strike on the airfield, damaging a large number of Me 262s and ground personnel. By the end of the month JG 7 had lost over half of its aircraft and been forced to move back.

Galland's JV 44 was suffering as well, although the flying personnel were boosted by the arrival of two of the Luftwaffe's top aces, Major Gerd Barkhorn and Major Günther Lützow.

The unit was in the position of having to scavenge for parts to make their aircraft serviceable and by the end of April they had twenty operationally ready aircraft and around 22 pilots. Then disaster struck the unit when two *Ketten* of six aircraft were scrambled to intercept heavy bombers on their way to Regensburg. The first three aircraft, led by Adolf Galland, took off, followed by the other three led by Johannes Steinhoff. As Steinhoff was about to lift off, his starboard wing dipped and dug into the ground, cartwheeling the aircraft down the runway. The Me 262 burst into a fireball but Steinhoff managed to extricate himself from the blazing wreckage. He suffered horrendous burns to his face and body and it took

The wreckage of Johannes Steinhoff's Me 262 in which he suffered horrendous burns – but survived.

Remains of Johannes Steinhoff's Me 262 engine.

Adolf Galland with Johannes Steinhoff.

many years of hospital treatment to bring him back to some form of normality. In 1969 a RAF plastic surgeon grafted skin from Steinhoff's arms to form new eyelids enabling him to close his eyes for the first time since the accident.

The end was in sight as the Allied forces moved closer and closer to Berlin and the bombers continued their devastating raids without hindrance. On 4 May 1945 the town of Salzburg surrendered to the Americans. In an effort to stop the Me 262s falling into the hands of the Russians, Galland, on orders from Bär, rode around the airfield on a tracked motorcycle, together with Hauptmann Walter Krupinski, and threw hand grenades into the Jumo engines. As word reached the Eastern Front that Adolf Hitler had shot himself and that Germany was surrendering, pilots jumped into their aircraft and headed back westwards to surrender to the British and American troops.

The war was over and the Germans defeated; but their contribution to the world of aviation was huge. They produced the world's first rocket planes, the first operational jet bomber and the first operational jet fighter. Although the Allies were close behind, the information gathered from captured German material accelerated the production of jet aircraft around the world. It was German technology that helped put the first man on the Moon and in so doing expanded man's horizons.

Side view of the Russian CY-9.

Four views of the Russian-built Cy-9, the identical version of the Me 262.

Part of the legacy – not of the Me 262 but of the German rocket and jet designers throughout this book; or rather an example of how the lessons had to be relearned. The Lockheed XFV-1 vertical takeoff fighter. This was not supposed to be an X plane but an aircraft built to fighter standards for the US Navy. It first flew on 16 June 1954. It never made a vertical ascent from a standing start and it was never put into production – and it's got a propeller! A reminder of the brilliance of some of those German designers, engineers and pilots and how advanced they were. For a slightly more relevant example of the jet legacy, see overleaf. (Lockheed)

Ryan's X-13 Vertijet 'tail sitter' was powered by a 10,000lb s.t. Rolls-Royce Avon RA-28-49 engine. The power-to-weight ratio ensured that this aircraft certainly could rise vertically. An ingenious sky-hook was employed on landing, whereby the pilot had to manoeuvre close enough in the vertical to a platform to get attached to a glorified fishing line and drawn in. Just two X-13s were built – both retired to museums. Another faltering step during the early jet age. (Ryan)

SPECIFICATIONS

Messerschmitt Me 328

Engines: Two Argus As 014 pulse jets of 660lb thrust each.
Wing span: 28ft. 2½in. (8.6m)
Length: 22ft. 5in. (6.8m)
Height: 5ft. 11in. (1.8m)
Weight Gross: 9,330lb. (4,232kg)
Weight Empty: 7,125lb. (3,232kg)
Maximum Speed: 434 mph. (698 km/h)
Service Ceiling: 22,300ft. (6,800m)
Maximum Range: 466 miles.
Armament: Two 2,205lb bombs.
Crew: One.

Heinkel He 280 V5

Engines: Two Heinkel HeS 8A of 1,650lb static thrust.
Wing span: 40 ft. (12.1m)
Length: 34ft. 1½in. (10.4m)
Height: 10ft. (3m)
Weight Gross: 9,482lb. (4,301kg)
Weight Empty: 7,073lb. (3,208kg)
Maximum Speed: 559 mph. (899 km/h)
Service Ceiling: 37,720ft. (11,497m)
Armament: Three 20mm MG.151 cannons.
Crew: One.

Heinkel He 178

Engines: One HeS-3 250kg static thrust.
Wing span: 23ft. 7in. (7.2m)
Wing Area: 98 sq. ft. (9.2 sq/m)
Length: 24ft. 6½in. (7.5m)
Height: 6ft. 10in. (2.1m)
Weight Gross: 4,396lb. (1,998 kg)
Weight Empty: 3,565lb. (1,620 kg)
Maximum Speed: 360 mph. (580 km/h)
Armament: None.
Crew: One.

Heinkel He 162A-2

Engines: One BMW 109-003E-1 of 1,760lb static thrust.
Wing span: 23ft. 7¼in. (7.2m)
Length: 29ft. 8½in. (9.0m)
Height: 8ft. 4½in. (2.5m)
Weight Gross: 5,490lb. (2,490 kg)
Weight Empty: 3,859lb. (1,750 kg)
Maximum Speed: 522 mph. (839 km/h)
Landing Speed: 103 mph. (165 km/h)
Service Ceiling: 36,080ft. (10,997m)
Armament: Two 20mm MG.151/20 cannons.
Crew: One.

Henschel Hs 132A

Engines: One BMW 003A-1 of 1,760lb static thrust.
Wing span: 23ft.7in. (7.1m)
Length: 29ft. 2½in. (8.9m)
Height: 10ft. 0½in. (3.0m)
Weight Gross: 7,496lb. (3,400 kg)
Weight Empty: 6,732lb. (3,053 kg)
Max. Speed: 485 mph. (780 km/h)
Service Ceiling: 33,820ft. (10,250m)
Armament: One 500lb bomb.
Crew: One.

Messerschmitt Me 163B Komet

Engines: One Walter 109-509A-0-1 rocket motor of 3,300lb static thrust.
Wing span: 30ft. 7in. (9.3m)
Wing Area: 211 sq. ft. (19.6m)
Length: 18ft. 8in. (5.7m)
Height: 8ft. 4in. (2.5m)
Weight Gross: 9,042lb. (4,110 kg)
Weight Empty: 4,191lb. (1,095 kg)
Maximum Speed: 596 mph. (960 km/h)
Landing Speed: 137 mph. (220 km/h)
Service Ceiling: 39,690ft. (12,100m)
Maximum Operational Range: 50 miles.
Armament: Two 30mm MG.108 cannons.
Crew: One.

Messerschmitt Me 262

Engines: Two Junkers Jumo 109-004B-1 of 1,980lb static thrust.
Wing span: 41ft. 1in. (12.5m)
Length: 34ft. 9½in. (10.6m)
Height: 12ft. 7in. (3.8m)
Weight Gross: 8,300lb. (9,808 kg)
Weight Empty: 10,820lb. (4,908 kg)
Maximum Speed: 487 mph. (783 km/h)
Service Ceiling: 31,090ft. (9,476m)
Armament: Single 50mm MK 214A cannon/ Two MK.108, MK 103 cannons and two MG 151/20 machine guns all mounted in the nose.
Crew: One.

Bachen Ba 349

Engines: One Walter HWK 109-509A-2 bi-fuel rocket of 3,740 thrust.
Wing span: 11ft. 9in. (3.6m)
Length: 19ft. 11in. (6.1m)
Height: 6ft. (1.8m)
Weight Gross: 4,850lb. (2,200 kg)
Weight Empty: 1,840lb. (834 kg)
Maximum Speed: 500 mph. (800 km/h)
Service Ceiling: 45,920ft. (14,000m)

Max. Operational Range: 25 miles. (40 km)
Armament: 33 R4M 55mm folding-fin rockets mounted in the nose.
Crew: One.

Fieseler Fi 103-IV

Engines: One Argus 109-014 pulse-jet 770lb thrust.
Wing span: 18ft. 9in. (5.7m)
Length: 26ft. 3in. (8.0m)
Height: 4ft. (1.2m)
Weight Gross:4,718lb. (2,140 kg)
Maximum Speed: 576 mph. (927 km/h)
Service Ceiling: 20,090ft. (6,123m)
Max. Operational Range: 23 miles. (37 km)
Armament: Aircraft was a flying bomb.
Crew: One.

Junkers Ju 287

Engines: Four Junkers Jumo 004B-1 Orkan axial-flow turbojets.
Wing span: 65ft. 11½in. (20.1m)
Length: 60ft. ½in. (18.3m)
Wing Area: 656.58 sq. ft. (58.3m)
Height: 12ft. 2ins. (36.8m)
Weight Gross: 44,100lb. (2,003 kg)
Weight Empty: 27,560lb. (12,512 kg)
Maximum Speed: 347 mph. (559 km/h)
Service Ceiling: 39,360ft. (12,000m)
Max. Operational Range: 985 miles. (1,585 km)
Armament: Twin 13-mm Mg 131 machine guns in remotely controlled tail barbette purely as a defensive weapon.
Crew: Two.

Arado 234 V1

Engines: Two Junkers Jumo 004A-O Orkan axial-flow turbojets.
Wing span: 46ft. 6in. (14.20m)
Length: 41ft. 3in. (12.58m)
Height: 12ft. 3ins. (3.75m)
Weight Empty: 10,473lb. (4750 kg)

Weight Loaded: 17,419lb. (7900kg)
Maximum Speed: 472mph. (760 km/h)
Service Ceiling: 30,090ft.
Max. Operational Range: 416 miles. (670km)
Armament: Four 20-mm MG 151/20 machine guns, two forward firing and two in remotely controlled rearward firing purely as a defensive weapon.
Crew: One.

TRANSLATIONS/GLOSSARY

The overall control of the *Luftwaffe* came under the *Reichsluftfahrtministerium* – RLM – (Reich Ministry of Aviation) who were responsible for both civil and military aviation. At the head of the organisation was Hermann Göring who was both *Oberbefehlshaber der Luftwaffe* (Commander-in-Chief of the Air Force) and *Reichsminister der Luftfahrt* (Minister of Aviation). With regard to the Luftwaffe the largest flying unit was a *Geschwader* (90 aircraft), which usually consisted of three *Gruppen* (30 aircraft each *Gruppe)*, which in turn was made up of three *Staffeln* (9–10 aircraft each).

The unit designations of the Luftwaffe were made up as follows and the translations as close to the literal as can be:

Kampf	Bomber
Kampfgeschwader	Bomber Unit
Jagd	Fighter
Jagdgeschwader	Fighter Unit.
Nachtjagd	Night Fighter
Nachtjagdgeschwader	Night Fighter Unit

A bomber unit would be known as a *Kampfgeschwader (KG)* this would be followed by the unit's number in Arabic numerals ie. *Kampfgeschwader 4.* This was the same for the *Jagdgeschwader (JG)* and *Nachtjagdgeschwader (NJG).* The individual *Gruppen* however was identified using Roman numerals ie. *IV/Nachtjagdgeschwader 3* which translates as the Fourth Group of Night Fighter Unit Three. The *Staffeln* however used the Arabic numerals to precede the *Geschwader* ie. *4/NJG 3* translated as 4th Staffel of Night Fighter Unit Three. Within the *Staffeln* aircraft flew in *Schwarmes* (Flights), *Ketten* (three) and *Rotten* (pairs).

1st Staffel Versuchstaffel der Oberkommando der Luftwaffe Luftwaffe High Command Trials Detachment

Deichselschlepp Towed Bomb

Deutsche Forschungsanstalt fur Segelflug (DFS) German Research Institute for Sailplane Flight

Deutches Versuchsanstaldt für Luftfahrt (DVL) German Aviation Experimental Establishment

Generalluftzeugmeister Director General of Luftwaffe Equipment

Generalluftzeugmeister-Besprechung Chief of Procurement and Supply Conference

Geschwader Unit comprising three or more *Gruppen*
Geschwaderstab Geschwader Staff
Gruppen Operational unit comprising three Staffeln
Heereswaffenamt Army Weapons Department
Jagdegerfaust Fighter Fist
Jagdgeschwader Fighter Unit
Kampfgeschwader Bomber Unit
Ketten Flight of three aircraft
Luftfahrtforschungsanstalt Aviation Research Establishment
Luftfahrtforschungsanstalt Wien Aviation Research Establishment, Vienna
Obfw – Oberfeldwebel Flight Sergeant
Objektschutz-Jägers Target Interceptors
Reichsluftfahrtministerium (RLM) German Air Ministry
Reichsforschungsführung Reich Research and Planning
Rotten Flight of two aircraft
Scheuschlepper Shy Tug
Schnellbomber Fast Bomber
Selbstopfermänner 'Self-sacrifice men', those who volunteered to man the Fi-103R Reichenberg, the V-1 in what would have amounted to suicide missions, had they been undertaken.
Sondertriebwerke Special Propulsion Systems
Staffeln Operational flying unit comprising nine aircraft
Sturmvogel Assault Bird
Volksjäger People's Fighter
Volksjäger-Erprobungskommando People's Fighter Evaluation Unit
Volkssturm Home Guard
Volkssturmstaffeln National Milita Formation
Zerstörergeschwader Long-range Unit

INDEX

A

Afrika Korps 160
Alt Loennewitz 130
Anklam 49
Antwerp 135, 171
Arado Ar 66 13
Arado Ar.234 127–140 passim
Arado Flugzeugwerke 127
Ardennes 134, 170
Argus Company 88, 95, 97–98, 122, 179, 182
Argus-Rohr 122
Aschersleben 116
Augsburg 27, 30, 36–38, 97, 147, 151–152, 161, 163–164
Axmann, Reichsjugendführer Artur 110–111

B

B-17 Flying Fortress 47, 54, 58, 61–62, 65–66, 68, 134, 166, 167–168, 170, 172–173
B-24 Liberator 61–62, 144, 160, 169–170
Bachem Ba.349 71–85 passim
Bachem, Erich 71, 80, 81
Bacon, Roger 7
Bad Zwischenahn 49, 51–52, 53
Baetcher, Major Hans-Georg 135
Battle of the Bulge 170
Baumbach, General Werner 89, 93
Bär, Oberstleutnant Hans 119
Beauvais, Heinrich 154
Belgium 134–135, 167
Benghazi 160
Berchtesgaden 161
Berlin 22, 27, 30, 37, 105, 110, 116, 123, 158, 167–168, 175
Bernardi, Mario de 13
Bernburg 116
Betz, Professor 110
Bigelow, Lieutenant Marvin AAF 172
Blohm & Voss 109–110, 159
Blume, Professor Walter 49, 127
BMW (Bavarian Motor Werke) 17, 106–107, 116, 120, 122, 125–126–130, 140, 144, 147–148, 150–154, 158, 180
B hner, Hauptman Otto 39
Boye, Hans 38
Brandis 56, 58, 62, 68–69, 144
Braun, Werner von 9, 12–13, 23, 71, 100
Braunegg, Oberstleutnant 170
Bremerhaven 22
Brieg 53
Brussels 135
Brüx 62
Burg 134
Busemann, Professor A 143

C

C-Stoff 24, 29, 33, 41, 43–44, 51, 67, 73, 88, 150
Campini-Caproni CC2 13, 103
Campini, Secondo 13
Caproni 9, 13, 103
Châteaudun, France 167
Chievres 133–134

Cologne 135
Croissant, Leutnant 138

D

Dachau 164
Daimler-Benz 22, 101
Dannenberg 89
De Havilland 46
Deichselschlepp 150, 165
Delmenhorst 56
Dendermonde 167
Deutsche Forschungsanstalt fur Segelflug (DFS – German Research Institute for Sailplane Flight) 95
Deutches Versuchsanstaldt für Luftfahrt (German Aviation Experimental Establishment 23, 143, 185
Dittmar, Heini 25, 27, 30–31, 35, 38, 42, 49, 55, 64
Dornberger, Walter 23
Dornier, Claude 159
Dornier Do.217E 87, 95–96, 98
Dülgen, Flugkapitän 146
Dunker, Dr. Erich 43
Duxford 167

E

Eder, Hauptmann Georg-Peter 170
Eissenlohr, Wolfram 105
El Alamein 154
Ente (Duck) 19
Entwicklungsamt (Development Office) 25
Erprobungskommando 39, 66, 119, 155, 163, 165, 166
Espenlaub 8, 19, 21
Espenlaub, Gottlob 8

F

Fairbanks, Squadron Leader David RAF 140
Felden, Hauptmann Hans 140
Fieseler Fi.103 90, 91, 92, 93
Flachs, Unteroffizier Hans 164
Focke-Achgelis 56
Focke-Wulf Fw.190. 87, 98, 106, 109, 119, 168
Frank, Leutnant Alfred 135
Francke, Dr. Ing 109
Frankfurt 21
Franz, Dr. Anselm 162
Führer 26, 111, 161, 162, 165, 166, 171, 173

G

Galland, General der Jagdflieger Adolf 12, 34, 37, 111, 156, 158, 161, 165, 166, 168, 170, 171, 172, 173, 175
Geschwaderstab 165, 170
Generalluftzeugmeister (Director general of Luftwaffe Equipment) 30, 153, 156, 185
Gilze Rijen 135
Gloster Meteor 139
Glushko, Valentin 19
Göbbels, Reichsmarschall Joseph 158
Goetz, Oberleutnant Horst 132, 133, 134
Gollob, Oberst Gordon 34, 52, 171, 172
Göring, Reichsmarschall Hermann 26, 37, 38, 102, 110, 111, 137, 158, 159, 160, 161, 162, 165, 166, 168, 171, 172, 185
Grosslau, Dr. Ing. Fritz 98
Güther, Siegfried 111

H

Habich (Hawk) 41
Halberstadt 116
Hanover 172
Hatchel, Leutnant Gustav 59, 60
Hawker Tempest 135, 140
Hays, Flying Officer M.R RAF 53
Heereswaffenamt (Army Weapons Department) 23
Heinkel, Professor Ernst 17, 26, 99, 100, 101, 102, 103, 104, 109, 111, 122, 147
Heinkel He.111 25, 82, 89, 104
Heinkel He.111H-6 107
Heinkel He.112 26, 100

Heinkel He.118 101
Heinkel He.162A 180
Heinkel He.162-V models 116
Heinkel He.176 26, 101
Heinkel He.178 5, 17, 99, 102, 180
Heinkel He.280 99, 103, 104, 105, 106, 179
Heinkel-Hirth 122, 126, 153
Heinkel-Nord 114
Henschel Flugzeugwerke 123, 124
Henschel Hs.132A 5, 123, 124, 125, 180
Hertel, Heinrich 26
Himmler, Reichsführer Heinrich 71, 110, 171
Hirth Motoren GmbH 105, 122, 123, 126, 153
Hitler, Adolf 26, 37, 38, 67, 88, 89, 110, 111, 113, 158, 159, 160, 161, 162, 165, 166, 181, 182
Hohmann, Bernard 38
Holland 134, 135
Holzbauer, Flugkapitän Siegfried 144
Hörsching, Austria 98

I

Ihlefeld, Oberst Herbert 119

J

Jacob Schweyer Segelflugzeugbau Company 95, 97, 98
Jagdgeschwader 33, 35, 52, 68, 119, 185
Junkers Jumo 98, 105, 106, 107, 119, 122, 127, 130, 144, 151, 152, 154, 162, 166, 181, 182
Junkers, Hugo 141, 147
Junkers Ju.88 98
Junkers Ju.287 5, 141, 142, 143, 144, 145, 146, 182
Juvincourt 132., 133, 134

K

Kamikaze 8
Kammhuber, Generalmajor Josef 111
Kammler, General der Waffen-SS Obergruppenführer 111, 173
Kampfgeschwader (Bomber Unit) 87, 135, 165, 166, 185
Kappus, Peter 140
Kelb, Leutnant Fritz 28, 48, 62
Keller, Generaloberst 110
Kensche, Heinz 89
Ketten 173, 185
Kettenfuhrer (Flight Leader) 37
Kiel 23, 27, 67
Kiel, Oberleutnant Johannes 36
Kircheim 82
Klein, Feldwebel Herbert 69
Kneymeyer, Oberst Siegfried 141, 143
Koller, General Karl 167
Komet 5, 9, 17, 19, 29, 31, 33, 34, 36, 39, 43, 44, 47, 49, 50, 51, 52, 53, 54, 55, 56, 59, 61, 65, 66, 67, 68, 69, 98, 144, 181
Korolyov, Sergei 19
Kranich (Crane) 41
Kreipe, Generalleutnant Werner 167
Kreutzberg, Obfw 167
Küchemann, Professor 110
Kummersdorf 23, 26
Kunos 163
Künzel, Walter 100

L

Lamm, Otto 51
Langer, Oberleutnant Herbert 36
Lauer, Obfw 167
Laverdiere, 2nd Lieutenant C.J. AAF 62
Lechfeld 27, 155, 156, 158, 165, 166, 167, 170
Leipheim 152, 163
Leipzig 53, 58, 59
Lennartz, Feldwebel 167
Leuan 62
Liege 135
Lille 53
Linder, Gerd 158, 162, 173

Lippisch, Alexander 8, 19, 21, 26, 27, 30, 31, 37, 38, 39
Lippisch-Latscherwagen 98
Lobban, Pilot Officer A.S. RAF 166
Lockheed XFV-1 177
Lucht, General Dipl.Ing 110, 154
Ludendorf 136, 138
Luftfahrtforschungsanstalt (Aviation Research Establishment) 39, 185
Luftwaffe (Air Force) 12, 26, 27, 30, 31, 35, 36, 37, 39, 49, 51, 52, 54
Lukesch, Hauptmann Diether 134, 135
Lüsser, Robert 102
Lützow, Major Gunther 173

M

Magdeburg 17, 103, 134
Marienehe 26, 100, 111, 114, 117
Martens-Fliegerschule (Martens Flying School) 21
Mauch, Hans A 104
Meister, Oberstleutant Wolf-Dietrich 165
Melsbroek 138
Meresburg-Leuna 58
Messerschmitt Bf,109 98, 134
Messerschmitt Me.163B 5, 9, 13, 17, 19, 24, 56, 57, 65, 82, 181
Messerschmitt Me.262 5, 14, 17, 106, 109, 147, 149, 151, 181
Messerschmitt Me.264 98
Messeschmitt Me.328A-2 97
Messerschmitt Me.328C 97
Messerschmitt, Willi 17, 27, 37, 38, 153, 154, 156, 159 160, 161, 162, 166
Mikhailov, Onisin 7
Milch, Generalfeldmarschall Erhard 26, 35, 37, 102, 103, 153, 154, 156, 160, 161
Mosquito 46, 49, 53, 166, 167, 168
Müller, Max Adolf 103, 105, 106
Münich 163, 166, 173
Münster-Handorf 134, 135
Murphy, Lieutenant-Colonel John AAF 62
Myers, Major Joe AAF 167

N

Natter 71, 74, 78, 82, 84, 85
Nebel, Rudolf 13
Nicolaus, Dipl.Ing 123
Nordhausen 116
Nowotny Kommando 153, 157, 168, 169
Nowotny, Major Walter 14, 168, 170

O

Oberth, Hermann 7, 9, 13
Objektschutz-Jägers (Target Interceptors) 56, 185
Oertzen, Oberleutnant Otto 39
Ohain, Pabst von 17, 99, 100, 101, 103, 105
Okha (Cherry Blossom) 122
Olejnik, Hauptmann Robert 39, 52, 56, 58, 66
Oldenburgh 49
Opel, Fritz von 19, 21, 22
Opitz, Rudolf 30, 31, 36, 37, 38, 39, 40, 43, 49, 50, 51, 52, 54, 55, 59, 62, 63
Oranienburg 116, 132
Osnabrück 134

P

P-38 Lightning 61, 167
P-47 Thunderbolt 50, 52, 167, 169, 173
P-51 Mustang 50, 59, 62, 68, 134, 168, 169, 170, 172
Pedace, Giovanni 13
Peenemünde 26, 27, 37, 38, 39, 46, 47, 49
Perschall, Franz 51
Peter, Flugkapitän Gotthold 116
Pohl, Professor 99
Pöhs, Oberleutnent Joschi 36, 37, 39, 50
Pulverhof IV 89

R

RAF (Royal Air Force) 101, 134, 139
Rak I 21, 22
Rak 3 8, 21
Rastenburg 173

Rebeski, Ing. 127
Reichenberg 5, 87, 88, 89, 92, 93, 185
Reichsluftfahrtministerium (RLM-German Air Ministry) 17, 27, 185
Reichsforschungsführung (Reich Research and Planning) 27
Reims 132, 133
Reitenbach, Ingnr. 105
Reitsch, Hanna 37, 42, 64, 89
Rhine 136, 138
Rolls-Royce Avon 178
Rolls-Royce Kestrel 101
Rommel, Erwin 154
Rostock-Marienehe 100
Rottens (Two-aircraft sorties) 173
Ruhland 62
Ryall, Leutnant Hartmut 62
Ryan X-13 Vertijet 178

S

Salamander 111
Salzburg 175
Sander, Alexander 19
Saur, Karl Otto 109
Saxony 116,130
Schäfer, Fritz 104, 153
Schelp, Helmut 104, 105, 106
Schenk, Major Wolfgang 165
Scheuschlepper (Shy Tug) 31, 50, 58, 64, 185
Schkeuditz 158
Schmidt, Kurt 164
Schmidt, Paul 98
Schnellbomber (Fast Bomber) 160
Schneider, Dr. 49
Schreiber, Leutnant Alfred 167
Schubert, Leutnant Siegfried 65,66
Schwarmes (Flights) 185
Schwärzler, Karl 111
Schweinfurt 160, 163
Seiler, Franz 154
Selbstopfermänner (self-sacrifice men) 89, 185
Selle, Flugkapitän 127
Siebert, Oberleutnant Lothar 76, 77, 79, 80, 82
Skorzeny, SS Colonel Otto 88
Sommer, Leutnant Erich 132, 134
Sondertriebwerke (Special Propulsion Systems) 25
Späte, Hauptmann Wolfgang 35, 36, 37, 39, 49, 51, 52, 68, 155, 156
Spatz (Sparrow) 111
Speer, Albert 37, 109, 110, 171
SRB (Solid Rocket Booster) 130
Stamer, Fritz 21
Steinhoff, Major Johannes 172, 173, 174, 175
Storch (Stork) 8, 19, 26, 49
Straznicky, Felwebel Herbert 61
Sturmvogel (Assault Bird) 165, 167
Stuttgart 82, 106, 116, 167
Stuttgart-Zuffenhausen 105

T

Thaler, Hauptman Toni 36, 39, 52, 53
Thierfelder, Hauptmann Werner 163
T-Stoff 33, 39, 41, 43, 45, 50, 51, 82
Tsiolkovsky, Konstatin 19

U

Udet, Ernst 26, 30, 31, 33, 35, 102, 103, 105, 153
Urseburg 116

V

Valier, Max 19, 21, 22, 23
Verran, Pilot Officer 135
Venlo 53
Vienna 39, 185
Vienna-Schwechat 110
Volkel 134
Volksjäger (Peoples Fighter) 109, 110, 111, 112
Volksjäger-Erprobungskommando (Peoples Fighter Evaluation Unit) 119

Volkssturm (Home Guard) 110
Volkssturmstaffeln 111
Voy, Karl 51
Vulcan Projekt 155

W

Wagner, Herbert 17
Waldsee 75
Wall, Flight Lieutenant A.E. RAF 166
Walter, Hellmuth 23, 25, 27, 33
Walter HWK (Hellmuth, Walter, Kiel) RII-203 27, 31, 33, 35, 39, 148, 158, 181
Walter TP-2 rocket engine 26
Warsitz, Erich 26, 100, 102, 103
Weimar 62
Weltensegler GmbH 19
Wermacht 43
Whittle, Frank 100, 101
Wiener-Neustadt 160
Wittmundhafen 53, 56
Wocke, Dipl.Ing 143
Wörndl, Feldwebel Alois 50

Z

Zacher, Flugkapitän 80
Zerstörergeschwader (Long-range Unit) 163
Ziegler, Leutant 67
Zindel, Dipl.Ing Ernst 143
Zueber, Flugkapitän 82